Debora Mahlke

A Scenario Tree-Based Decomposition for Solving Multistage Stochastic Programs

VIEWEG+TEUBNER RESEARCH

Stochastic Programming

Editor:

Prof. Dr. Rüdiger Schultz

Uncertainty is a prevailing issue in a growing number of optimization problems in science, engineering, and economics. Stochastic programming offers a flexible methodology for mathematical optimization problems involving uncertain parameters for which probabilistic information is available. This covers model formulation, model analysis, numerical solution methods, and practical implementations. The series "Stochastic Programming" presents original research from this range of topics.

Debora Mahlke

A Scenario Tree-Based Decomposition for Solving Multistage Stochastic Programs

With Application in Energy Production

VIEWEG+TEUBNER RESEARCH

Bibliographic information published by the Deutsche Nationalbibliothek
The Deutsche Nationalbibliothek lists this publication in the Deutsche Nationalbibliografie; detailed bibliographic data are available in the Internet at http://dnb.d-nb.de.

Dissertation Technische Universität Darmstadt, 2010

D 17

1st Edition 2011

Editorial Office: Ute Wrasmann | Anita Wilke

Vieweg+Teubner Verlag is a brand of Springer Fachmedien.
Springer Fachmedien is part of Springer Science+Business Media.
www.viewegteubner.de

Cover design: KünkelLopka Medienentwicklung, Heidelberg
Printing company: STRAUSS GMBH, Mörlenbach
Printed on acid-free paper
Printed in Germany

ISBN 978-3-8348-1409-8

Acknowledgments

First of all, I would like to thank all those people who have helped and supported me during the completion of this work.

Especially, I would like to express my gratitude to my advisor Professor Alexander Martin for giving me the opportunity to carry out my research work in his group. Besides his continuous support and guidance, he encouraged me to follow my own ideas and enabled me to attend many conferences where I could present my work. Furthermore, I am grateful to my co-referee Professor Rüdiger Schultz for providing a second opinion and for the motivating support of my research in the field of Stochastic Optimization.

Special thanks go to my colleague and friend Andrea Zelmer for the numerous motivating discussions and the intensive collaboration which I have enjoyed from the first day on. I also like to thank her and Ute Günther for the valuable support during the final completion of this thesis and the inspiring and amicable time we shared in our office.

Additionally, I am grateful to all colleagues from my working group for the cooperative spirit and the pleasant working atmosphere. In particular, I thank Thorsten Gellermann, Ute Günther, Wolfgang Hess, Henning Homfeld, Lars Schewe, Stefan Vigerske, Andrea Zelmer, and Nicole Ziems for proofreading parts of this thesis. Likewise, I am grateful to all members of the BMBF network in particular to Alexa Epe, Oliver Woll, and Stefan Vigerske for the successful cooperation.

Last but not least, I like to thank my family and especially Andreas for the loving support, continuous encouragement, and understanding.

Debora Mahlke

Abstract

This thesis is concerned with the development and implementation of an optimization method for the solution of multistage stochastic mixed-integer programs arising in energy production. Motivated by the strong increase in electricity produced from wind energy, we investigate the question of how energy storages may contribute to integrate the strongly fluctuating wind power into the electric power network. In order to study the economics of energy storages, we consider a power generation system which consists of conventional power plants, different types of energy storages, and an off-shore wind park which supplies a region of certain dimension with electrical energy. On this basis, we aim at optimizing the commitment of the facilities over several days minimizing the overall costs.

We formulate the problem as a mixed-integer optimization program concentrating on the combinatorial and stochastic aspects. The nonlinearities arising from partial load efficiencies of the units are approximated by piece-wise linear functions. In order to account for the uncertainty regarding the fluctuations of the available wind power and of the prices for electricity purchased on the spot market, we describe the affected data via a scenario tree. Altogether, we obtain a stochastic multistage mixed-integer problem (SMIP) of high complexity whose solution is algorithmically and computationally challenging.

The main focus of this thesis is on the development of a scenario tree-based decomposition approach combined with a branch-and-bound method (*SD-BB*) for the solution of the SMIP described above. This novel method relies on the decomposition of the original formulation into several subproblems based on the splitting of the scenario tree into subtrees. Using a branch-and-bound framework which we extend by Lagrangian relaxation, we can solve the problem to global optimality. In order to support the solution process, we investigate the polyhedral substructure which results from the description of switching processes in a scenario tree formulation yielding a complete linear description of the polytope. Furthermore, we develop an approximate-and-fix heuristic which generates feasible solutions of the problem with low computational effort. In the sequel, we specify the implementation of the *SD-BB* algorithm where we exploit problem-specific as well as algorithmic properties to make the method successful. Although our algorithm has originally been devised for a specific kind of problem, the general framework can be applied for the solution of a wide range of related problems. Finally, we evaluate the performance of the developed methods based on a set of test instances close to reality. Applying the *SD-BB* algorithm, we consider instances with a time horizon ranging from six hours up to several days using a time discretization of 15 minutes.

Contents

List of Figures

List of Tables

Chapter 1

Overview

Power generation based on renewable energy sources plays an important role in the development of a sustainable and environmentally friendly generation of energy, motivated by the finite nature of fossil energy sources and environmental pollution. In particular, wind energy is considered to be most promising to provide a substantial part of the electrical energy supply. But due to the fluctuating behavior of power production from renewable energies, especially caused by wind power production, new challenges are posed to the structure of power generation systems. In this context, we approach the question of how energy storages and flexible generation units may contribute to decouple fluctuating supply and demand, yielding a sustainable and cost efficient energy production. To this end, the problem is formulated as an optimization model including combinatorial, nonlinear, and stochastic aspects. By approximating the nonlinearities, we receive a stochastic multistage mixed-integer program. The aim of this thesis is the development of a solution algorithm which is capable to solve test instances sufficiently large to provide reliable results. This is accomplished by developing a decomposition approach based on splitting the corresponding scenario tree, enhanced by mixed integer programming techniques, such as primal methods and cutting plane generation.

In detail, the thesis is structured as follows. In Chapter 2, we give an introduction to the power generation problem, which arises when large amounts of fluctuating energy are fed into the public supply network. In this context, the focus is on the potential of energy storages in order to decouple supply and demand. Next to a description of the basic technical characteristics of the facilities considered in the generation system, a survey on related literature is given regarding modeling and solution approaches.

Chapter 3 contains the mathematical modeling of the power generation problem. In the first part, a deterministic model is presented assuming all data to be known in advance. With the aim of a realistic description, partial load efficiencies of the facilities are taken into account leading to the integration of nonlinear functions into the model. In order to handle the resulting mixed integer nonlinear problem, an approximation of the nonlinearities by piecewise linear functions is described. In the second part of this chapter, we extend the model towards the inclusion of uncertainty concerning the amount of wind power available and the market prices for electricity. Using a scenario tree approach to describe the evolution of the uncertain data, we formulate a multistage stochastic mixed-integer problem.

Chapter 4 addresses the investigation of polyhedral substructures of the problem. In particular, we investigate the facial structure of the polytope arising from the description of switching processes with minimum running time and minimum down time restrictions in a scenario tree formulation. Based on the results for the deterministic case, we derive a complete linear description of the polytope occurring within the stochastic formulation. Using these inequalities as cutting planes we incorporate them in the solution process of the problem described above.

The focus of Chapter 5 is on the generation of good feasible solutions based on the idea of relax-and-fix. With regard to the deterministic problem formulation, an adapted rolling horizon algorithm is presented, where the relaxation of the integrality conditions is enhanced by problem specific approximation schemes. Assuming the original problem to be feasible, we investigate the possibility of running into infeasible subproblems. In this context, we show that the algorithm terminates with a feasible solution of the entire problem, imposing certain conditions on the input data and the approximation scheme. Subsequently, we provide an adaptation to the stochastic problem by extending the generation of the subproblems, approximation schemes, and feasibility results yielding an approximate-and-fix heuristic.

One crucial point of this thesis concerns the development of a novel solution approach to the stochastic power generation problem from above which is presented in Chapter 6. We reformulate the original problem by decomposing it into several subproblems coupled by few coupling constraints which is based on the splitting of the scenario tree into subtrees. In order to determine global optimal solutions of the problem, we integrate this approach into a branch-and-bound framework called *SD-BB* (scenario tree based decomposition combined with branch-and-bound). Furthermore, we extend this method by applying Lagrangian relaxation in order to generate tighter lower bounds of the optimal solution value.

In Chapter 7, we describe the implementation of the *SD-BB* algorithm mentioned above. First, we focus on its initialization phase whose core comprises the splitting of the scenario tree into several subtrees where we present a polynomial time algorithm. Furthermore, we discuss and specify suitable branching techniques for *SD-BB* focusing on variable selection rules and the determination of branching points in case of continuous variables. Finally, we address the computation of dual bounds as well as the determination of feasible solutions, exploiting the special structure of the problem at hand.

In order to evaluate the performance of the developed methods various test runs are performed which are summarized in Chapter 8. Besides the presentation of the numerical effects applying the developed separation algorithm and the primal heuristics, we focus on the computational investigation of the *SD-BB* algorithm. By applying the algorithm to various instances scaled with respect to the basic properties of the *S-OPGen* problem and comparing the results with the commercial solver CPLEX the performance of the algorithm is investigated. Finally, we complete this thesis in Chapter 9 with a conclusion and suggestions for further improvements and investigations.

Chapter 2

An Energy Production Problem

The focus of this chapter is on the description of the power generation problem constituting the energy economical application of this thesis. In Section 2.1, we start with an introduction to the energy economical background exposing the issues arising from an increasing participation of regenerative energy to the electric energy supply. Subsequently, we give a description of the power supply problem studied in this thesis focusing on the application of energy storages in order to decouple fluctuating supply and demand in Section 2.2. In the power generation system, different power plants and energy storages are considered whose basic technical characteristics are presented in Section 2.3. Finally, in Section 2.4 we provide a survey on related literature concerning modeling and solution approaches of the problem.

2.1 Introduction

A sustainable, competitive and secure generation of electrical energy is a major aspect within the economic development of a country. To this day, fossil energy sources combined with nuclear power still dominate the energy mix of the electrical power generation in Germany. For instance, in 2007 the electrical supply was basically provided by using lignite (28 %), hard coal (28 %), and nuclear power (25 %), see [BMW]. As Germany is relatively poor in fossil energy resources, an essential part of the electric load is satisfied by imported energy. Particularly, uranium and natural gas mostly come from foreign suppliers as well as a growing part of hard coal resulting in a high import dependence. Energy from renewable resources is predominantly available in Germany and therefore, contributes to a higher energy independence. Additionally, motivated by limited fossil resources and environmental pollution, renewable energies continue to gain in significance.

Consequently, the power generation based on renewable energies was more than doubled from 2000 to 2007 to a level of 87 TWh, which is about 14 % of Germany's entire electric energy consumption, see [Koh08]. This development was significantly supported by the German Government, which legislated several regulations yielding a higher efficiency and a sustainable usage of energy. Particularly in 2000, the renewable energy law (EEG) was passed, concerning the priority for the feed-in of renewable energies and ensuring legally regulated payment rates, see [EEG00]. Due to the good development in this sector, the objective of providing at least 12.5 % of the electric energy consumption by renewable energies was already achieved in 2007. In 2008, an amendment to the EEG even increased the long term perspective to 30 % in 2020. Renewable energies comprise all kinds of energy produced from natural resources such as wind energy, solar radiant energy, geothermal heat, energy from biomass, hydroelectricity, and energy from tide and waves. They are characterized by a constant availability in the course of time by human standards, in contrast to the limited availability of fossil fuels like coal or natural gas. This means that renewable energies are derived from almost inexhaustible energy sources replenishing naturally.

A substantial contribution to the development of electricity generation from renewable sources has been made by wind power production. In 2001, the amount of wind energy was about 13 TWh, which satisfied already more than 2.5 % of the electrical energy load. Until now, a dynamic increase in produced wind energy can be observed, providing more than 39 TWh in 2007 (6.4 %), see [Koh08]. With regard to further expansions of renewables, wind is considered to be the most promising for providing a significant part of the electrical energy in Germany. Technically, wind energy production is well developed and feeding in large amounts of power is well experienced. But due to the permanently changing wind situation, the power generation is strongly fluctuating and additionally lacks reliability. Assuming a moderate development of the German grid infrastructure, a maximal participation of 20 to 25 % of wind energy of the electric load is assessed, according to the DENA study [Den05] in 2005. This motivates the search for further possibilities to increase the amount of regenerative energy supply.

2.2 Problem Description

Within the scope of a raising participation of wind power in electric power generation, subsequently we discuss the potential of energy storages to contribute to decouple supply and demand in order to achieve a cost-efficient power generation.

As described above, the fluctuating behavior of power production from wind energy poses new challenges to the structure of power generation systems. The increasing feed-in of fluctuating power into the electricity grid influences the operating requirements of the conventional power plants leading to a rising participation to the regulating energy. This means that the load to be satisfied by the plants is no longer only governed by the consumers' demand, but also by the wind power supply. The additional necessity of regulation affects the efficiency of power plants, as the generation efficiency strongly depends on the current production level.

A possibility of further increases in wind energy is the commitment of energy storages in the generation system. Although to a limited extend, energy storages are capable to convert and store surplus energy when produced and to supply energy in times of peak demand. Thus, transforming base load capacity to peak load capacity they can contribute to prevent partial load operation of conventional power plants. Additionally, they offer the ability to provide reserve energy which increases the security of energy supplies. Among others the following technologies are presently available: pumped hydro storage, compressed air energy storage, hydrogen fuel cell, batteries, flywheel energy storage, and supercapacitators. Today the most economical method is the pumped hydro storage, being used over decades for providing energy in times of high demand. In a power generation system, the prices for electricity have a major impact on the decision about the commitment of energy storages. Due to the liberalization of the German energy market, the realized market price for electricity has to be taken into account which is hardly predictable. Under consideration of theses facts, we face the question of how energy storages can contribute to decouple demand and supply, if large amounts of fluctuating regenerative energy are integrated into a power generation network.

To this end, we consider a power generation system which consists of conventional power plants and a wind park, supplying a region of a certain dimension with energy. In order to balance supply and demand, energy storages are integrated into the system. Additionally, we include the possibility of procuring power from an external supply network to which our system is connected. On this basis, the operation of these facilities is optimized, aiming at a cost-efficient generation of energy. In order to receive reliable results regarding the possible application of energy storages, planning horizons up to one week are addressed. Additionally, a time discretization of 15 minutes is desirable, permitting the consideration of partial load efficiency and yielding a realistic description of the power generation system.

By means of the results obtained from the optimization, the question about the potential of energy storages should be answered. In detail, the power supply system is characterized as follows.

Concerning the power generation, the main characteristic of the problem is the integration of wind power. As the power generation from wind energy strongly depends on meteorological conditions, uncertainty about the amount of power available has to be taken into account. As a consequence fluctuations are no longer only induced by consumers but also by generation units.

As indicated above, fossil fuel power plants are basically used for continuous operation but are also capable to provide regulating power adapting their generation level. The different types of plants differ in the fuel they burn such as coal, natural gas or petroleum. Each plant has a minimal load to ensure a stable combustion and a maximal load also called capacity. In order to control thermal stress of the units, the power gradient is limited. For the same reason, the starting time has to be stretched over a certain time period. A basic characteristic of a plant is the efficiency, describing the ratio between the useful output and the input. The efficiency of a power plant strongly depends on the current production level, i.e., the operation in partial load leads to a reduced generation efficiency.

The major task of an energy storage is to convert and store energy in times of low demand and to provide electric energy in peak times. But only the conversion of electric energy into other forms of energy allows the storage of large amounts of electricity over longer periods. As in the case of power plants, energy conversion involves energy losses, described by the conversion efficiency for charging and discharging. Furthermore, energy storages are characterized by their capacity and their flexibility regarding charging and discharging operations. For our investigations, we consider selected types of power plants and energy storages varying in their technological and economical characteristics. These technologies are specified in Section 2.3.

Next to the technical restrictions of the facilities described above, the major requirement in a power system is the balance of supply and demand at any time. In order to guarantee security of supply, variations of generated or consumed power need to be compensated as fast as possible. The objective of this problem concerns the minimization of the overall costs caused by the facilities and the imported electricity. In this context, the uncertainty about the realized market price for electricity has to be taken into account in order to provide reliable results.

Altogether, optimizing the operation of power plants and energy storages under the consideration of uncertainty, switching processes and partial load behavior for a planning horizon of one week poses a great challenge from the computational point of view, as integer as well as stochastic aspects are comprised in one model. This motivates the following studies presented in this thesis in order to be able to consider planning horizons large enough to receive meaningful results.

2.3 Technical Background

In this section, we present the basic technical and economical characteristics of the selected facilities considered in the power supply system. We start with the description of fossil-fuel power plants, followed by the characterization of energy storages. Within this scope, we describe the most common technologies which are suitable for our applications. With respect to power generation, we restrict our consideration to the following fossil-fuel power plants complying with the future German power mix proposed by [Enq02].

2.3.1 Fossil-Fuel Power Plants

In order to generate electrical energy, fossil-fuel power plants burn fossils such as lignite, hard coal or natural gas. They are commonly used to cover the base and medium load, except for gas turbines which are capable to satisfy peak demand. As lignite-fired power plants are characterized by slow operational changes and are not suitable for providing regulation energy, in the following we focus on the description of hard-coal and gas-turbine power plants.

Hard coal fired power plants belong to the thermal power stations, basically used to cover medium load. Technically, in a coal power plant the generation of electricity starts by pulverizing the coal which subsequently is burned in a furnace in order to heat water turning into steam. The pressurized steam is passed to a steam turbine, which spins a generator producing electricity. Finally, the water is condensed in a condenser and the process starts again. Coal power plants are characterized by a relatively high efficiency of about 38 to 46 %, see [Den05]. Additionally, these plants provide a good partial load efficiency, which means that for instance only 4 % of the maximal generation efficiency is lost if half of the capacity is used. But on the other hand their flexibility regarding operational changes is limited due to a low power gradient of about 4 to 8 % per minute. Concerning the start-up lag time, approximately four hours are needed if the plant has been turned off for at least eight hours, see [Gri07]. As a start-up also causes additional costs, a hard coal fired power plant is only suitable to a limited extent for contributing to balance supply and demand.

An established option to cope with fluctuating supply and demand is the use of *gas-turbine power plants* providing the ability to be turned on and off within minutes. These plants require natural gas as fuel using gas turbines as prime mover. Technically, a gas turbine mixes compressed air with gas and subsequently ignites the mixture under high pressure. The combustion develops hot gas which is directed through the turbine blades spinning the shaft. The resulting mechanical energy is used to drive the generator producing electrical energy. Using a significant part of the mechanical energy to run the compressor, a gas turbine shows a lower efficiency of about 39 % compared to a coal power plant. Additionally, operating a gas-turbine power plant in partial load strongly decreases its efficiency by up to 20 % of the maximal efficiency, see [Gri07]. The great advantage of a gas turbine lies in its flexible operation, providing a power gradient of 10 to 25 % per minute. Showing start up and shut down times of a couple of minutes, these power plants are ideally suited for covering peak load.

2.3.2 Energy Storages

The application of energy storages is determined by their capacity and time scale. With the aim of decoupling fluctuating supply from demand, only large scale options come into consideration. Hence, we restrict this section to the description of pumped hydro storage plants and compressed air energy storages, capable of providing significant reserve services. In the following, these facilities are described in detail.

Pumped hydro storage is the most common technology of storing energy in order to compensate peak loads in an energy system. In times of low demand, surplus power is used to pump water from a lower reservoir to a reservoir of higher elevation. When required, this upper reservoir provides the possibility to release water through turbines to generate electricity. Thus, in times of peak demand, electrical energy can be supplied within minutes. In general, water losses caused by evaporation or seepage are negligibly small so that the storage efficiency is basically determined by the pumps and turbines. As both exhibit a high efficiency per unit, the overall efficiency achieves values of up to 80 %, see [LG04]. Germany has about 5 GW of pumped storage capacity, where the largest storage in Goldisthal was put into operation in 2003, capable of providing 1060 MW for a maximal duration of eight hours when full, see [Gri07]. Requiring a certain vertical height, pumped storages find good conditions especially in mountainous parts of the country. Using flooded mines or caverns, underground pumped storages are also possible.

An alternative possibility of storing energy is provided by *the compressed air energy storage*, where air is compressed and stored under pressure to be used later for energy generation. The storage consists of a compressor and a turbine unit. During periods of low demand, the electrically driven compressor squeezes air with a pressure up to 100 bar into underground caverns, for instance old salt caverns or abandoned mines. In order to generate power in times of peak load, the compressed air together with additional natural gas is used to drive a gas turbine generating energy again. The compressed air substitutes the compressor of a conventional gas turbine, which normally requires about two thirds of the generated energy. Thus, the generated mechanical energy can be used completely to run the generator. As the compressed air must be reheated before the expansion in the turbine, the overall efficiency is degraded to about 40 %. Presently adiabatic storages are under development, storing the heat resulting during air compression as well as the pressurized air. Rendering co-firing unnecessary, here efficiencies up to 70 % are reached, see [Den05]. Actually, there are two diabatic storages, one in Huntdorf, Germany, with a maximal capacity of 240 MW for two hours and one in McIntosh, Alabama, with a maximal capacity of 100 MW for 26 hours, see [Gri07]. As for pumped storages, the application of compressed air storages is also restricted to the availability of suitable environmental conditions as for instance underground caverns. The creation of man-made tanks is currently under development.

2.4 Related Literature

In the literature a growing number of contributions regarding production planning problems in power generation can be found. A wide range of modeling aspects are studied followed by the proposition of solution approaches. In this section, we present several publications related to the problem described above and discuss their relevance for our studies. Thereby, we focus on the literature which concerns the same field of interest as this thesis. For contributions addressing multistage stochastic mixed-integer problems in general we refer to Chapter 6.

A number of papers focus on the deterministic formulation of the problem, neglecting uncertainty regarding the load or wind power available. With regard to the unit commitment problem in power generation, the most common approach of solving the resulting mixed-integer program is the use of Lagrangian relaxation, see [MK77, BLSP83, Bal95, GMN+99]. In particular, [MK77] presented one of the first versions of this solution approach to power scheduling, relaxing all constraints which couple power units. Thus, the problem decomposes into single unit subproblems, which can be solved by dynamic programming. In order to solve the non-differential Lagrangian dual, a subgradient method is used. In [BLSP83], an enhancement regarding the solution of the dynamic program and the Lagrangian dual is provided. [Bal95] presented a generalized version of the problem, including minimum-up and minimum-down times, ramp constraints and transmission. Also further approaches using LP-based branch-and-bound are considered in the literature. A comparison of primal and dual methods can be found in [GMN+99], where the dual method is based on Lagrangian relaxation.

Regarding heuristic approaches, [Lee88] presented an algorithm which uses priority lists in order to rank the thermal units yielding a good commitment schedule. [HH97] proposed a genetic based neural networks algorithm providing good computational results. Also simulated annealing and genetic algorithms are implemented being flexible in handling additional constraints, see [ZG90] and [KBP96], respectively. The drawback of heuristic approaches lies in the absence of quality certificates. For further references concerning deterministic approaches in power generation, we refer to the literature survey published by [SK98].

More recently various papers appeared concerning the stochastic formulation of the problem. As for the deterministic formulation, the use of Lagrangian relaxation is very popular. Presently there are two decomposition approaches for multistage stochastic mixed-integer programs mainly discussed in the literature: the scenario decomposition, see [LW96, TBL96, CS98, CS99] and single unit decomposition, see [NR00, GKKN+02]. In the latter approach, the authors extend the Lagrangian relaxation approach to stochastic power generation problems, where uncertainty is approximated by a finite number of scenarios yielding a block structured mixed-integer program. Relaxing all coupling constraints between the units, the problem decomposes into stochastic single unit subproblems. In contrast to the model presented in this thesis, certain aspects as e.g. partial load efficiency are neglected, allowing a fast solution of the resulting subproblems. In [NR00] and [GKKN+02] the Lagrangian dual is solved by using a proximal bundle method combined with a Lagrange heuristic to produce near optimal solutions. The scenario decomposition approach presented in [CS98] and

[CS99] is based on Lagrangian relaxation with respect to non-anticipativity constraints combined with branch-and-bound to ensure convergence. Presenting computational results for two stage problems, the authors state the applicability to multistage problems as well. In [TBL96], the problem is also decomposed into finitely many deterministic subproblems applying progressive hedging. Finally, [LW96] uses a combination of progressive hedging and tabu search to solve the multistage problem. Note that using progressive hedging, global optimality can only be guaranteed for the convex case.

Finally, we refer to the publications [EKR⁺07] and [EKR⁺09]. The models presented there are closely related to the problem described above as these papers as well as this thesis arose within the scope of the same project. Neglecting partial load efficiency and combinatorial aspects, the authors focus on the representation of uncertainty via recombining scenario trees. For the solution, they developed a modified Nested Benders decomposition approach which is described in [KV07].

Summarizing, none of the presented publications completely covers the problem we are focusing on. Some modeling approaches are restricted to the deterministic formulation, i.e., uncertainty about wind power or regenerative energy supply in general is neglected. Further on, either the combinatorial aspects or the nonlinearities regarding the efficiency of the plants are not taken into account. To the best of our knowledge, no contributions of current literature consider a model which combines uncertainty, switching processes and partial load efficiency with a detailed description of the technical characteristics of the units, as done in this thesis. This means that here we focus on the solution of a stochastic multistage mixed-integer problem guaranteeing global optimality in dependence of the approximation of the nonlinear functions.

Chapter 3

Mathematical Modeling

The aim of this study is to analyze the potential of energy storages within a power generating system, including fluctuating energy supply. In this chapter we describe the mathematical modeling of the problem, taking technical as well as economical aspects into account.

The problem considered here is to optimize the energy supply of a city, where energy is generated by conventional power plants, purchased on the spot market or obtained from wind energy. Due to the growing proportion of energy produced from renewable energy sources, not only fluctuating demand has to be taken into account but also fluctuations of the power supply. In this system, energy storages are used to decouple supply and demand, achieving a better capacity utilization and a higher efficiency of the power plants. As we aim for a realistic formulation, the consideration of partial load efficiency of the facilities is preferable. In particular, certain characteristic curves are assigned to each facility, representing its operational behavior. As the resulting efficiency function normally behaves nonlinearly, we follow the approach of a piecewise linear approximation of these nonlinearities in order to handle the resulting complex problem.

Within a power generating system, the strongly fluctuating wind energy supply plays an important role. Also the variations of the market price of electricity significantly affect the operations of the energy storages. In a first step, a known profiles of the procured wind energy and electricity prices are assumed, which lead to a deterministic problem formulation. As the wind power production strongly depends on meteorological conditions, the uncertainty concerning the amount of wind energy should be taken into account. Also the market price for electricity is unpredictable and varies over time. In order to generate solutions that hedge against this uncertainty, we extend

Table 3.1: Sets

$\mathcal{T}$	set of all time steps of the planning horizon
$\mathcal{I}$	set of power plants
$\mathcal{J}$	set of energy storages
$\mathcal{K}_j$	set of charging units of storage $j \in \mathcal{J}$
$\mathcal{L}_j$	set of discharging units of storage $j \in \mathcal{J}$

the model to a multistage stochastic mixed-integer problem, where the uncertainty is represented by a multivariate stochastic process. As schedulers may give a reliable forecast of the consumers' demand concerning periods of one day or even one week, we assume that the corresponding load is given as a profile.

In the following section, we start with the introduction of the basic sets, parameters and variables, which are necessary to set up the model. On this basis, the deterministic model is formulated in Section 3.1, yielding a mixed-integer nonlinear problem (MINLP). Here, the problem description includes a generic description of power plants and energy storages, whereas a specification of selected types of facilities can be found in Section 8.1.1, where the problem instances for the test runs are described in detail. The linearization of the nonlinear components presented in 3.1.5 completes the description of the deterministic model which is formulated as a mixed-integer linear problem (MILP). Based on this formulation, the stochastic model is developed in Section 3.2. Introducing the basic concept for representing uncertainty via a scenario tree, we formulate the multistage stochastic problem as a large-scale MILP.

3.1 Deterministic Model

In this section, we present the deterministic model assuming that all values of the parameters are previously known. In particular, the wind power generation and the electricity prices are given by a profile, providing the basis for a deterministic formulation.

3.1.1 Sets and Parameters

In the following, we will introduce the necessary sets and parameters of the model. Table 3.1 shows a list of all sets and Table 3.2 gives an overview over the major parameters. For the optimization, we are interested in the con-

Table 3.2: Parameters

τ	time steps per hour	[1/h]
δ_t	consumers demand in period $t \in \mathcal{T}$	[MW]
ω_t	wind power supply in period $t \in \mathcal{T}$	[MW]
γ_t^{imp}	import costs per MWh in period $t \in \mathcal{T}$	[€/MWh]
p_i^{min}	minimum production level of plant $i \in \mathcal{I}$	[MW]
p_i^{max}	maximum production level of plant $i \in \mathcal{I}$	[MW]
$\bar{p}_{i1}$	initial production level of plant $i \in \mathcal{I}$	[MW]
Δ_i^{max}	maximum power gradient of plant $i \in \mathcal{I}$	[MW]
$\bar{z}_{i1}$	initial operational state of plant $i \in \mathcal{I}$	[1]
θ_i^{up}	minimum running time of plant $i \in \mathcal{I}$	[1]
θ_i^{down}	minimum down time of plant $i \in \mathcal{I}$	[1]
γ_i^{fuel}	fuel costs per MWh of plant $i \in \mathcal{I}$	[€/MWh]
γ_i^{var}	variable costs per MWh of plant $i \in \mathcal{I}$	[€/MWh]
γ_i^{up}	start-up costs of plant $i \in \mathcal{I}$	[€]
s_j^{min}	minimum storage level of storage $j \in \mathcal{J}$	[MWh]
s_j^{max}	maximum storage level of storage $j \in \mathcal{J}$	[MWh]
$\bar{s}_{j1}$	initial storage level of storage $j \in \mathcal{J}$	[MWh]
$\bar{s}_{jT}$	final storage level of storage $j \in \mathcal{J}$	[MWh]
$s_k^{in,min}$	minimum charging of unit $k \in \mathcal{K}_j$ and $j \in \mathcal{J}$	[MW]
$s_k^{in,max}$	maximum charging of unit $k \in \mathcal{K}_j$ and $j \in \mathcal{J}$	[MW]
$\bar{z}_{k1}^{in}$	initial charging state of unit $k \in \mathcal{K}_j$ and $j \in \mathcal{J}$	[1]
$s_l^{out,min}$	minimum discharging of unit $l \in \mathcal{L}_j$ and $j \in \mathcal{J}$	[MW]
$s_l^{out,max}$	maximum discharging of unit $l \in \mathcal{L}_j$ and $j \in \mathcal{J}$	[MW]
$\bar{z}_{l1}^{out}$	initial discharging state of unit $l \in \mathcal{L}_j$ and $j \in \mathcal{J}$	[1]
α_j^{in}	start-up energy for charging of storage $j \in \mathcal{J}$	[MWh]
α_j^{out}	start-up energy for discharging of storage $j \in \mathcal{J}$	[MWh]
$\gamma_j^{in,up}$	start-up costs for charging units of storage $j \in \mathcal{J}$	[€]
$\gamma_j^{out,up}$	start-up costs for discharging units of storage $j \in \mathcal{J}$	[€]
γ_j^{fuel}	fuel costs per MWh of storage $j \in \mathcal{J}$	[€/MWh]

sideration of a predefined planning horizon. This span of time is discretized into subintervals, taking the restricted availability of, for example, load or wind data into account and making the problem tractable. Subdividing the horizon into T time periods, we obtain the set $\mathcal{T} = \{1, \ldots, T\}$, where the index $t \in \mathcal{T}$ represents a time period of the planning horizon.

In the model, the produced or consumed power is measured in MWh per hour. As we are also interested in time discretizations of less than one hour, we introduce the parameter τ, indicating the number of time periods per hour. For each time period $t \in \mathcal{T}$, the parameter δ_t denotes the consumer's demand. In order to cover the arising load, the available wind power can be used, represented by ω_t. A further possibility is to import power from an external power supply network, where the costs per unit of imported energy are given by γ_t^{imp}. Note that δ_t, w_t, and γ_t^{imp} vary over the time periods. Finally, conventional power plants may be used for power production, which are characterized by the following parameters.

Conventional Power Plants

Let $\mathcal{I}$ be the set of all power plants considered in the generation system. A power plant $i \in \mathcal{I}$ is a controllable unit, which produces power within certain bounds. By p_i^{max}, we refer to the maximum power production level, also called *installed capacity*, and by p_i^{min} to the minimum power production level. A further technical parameter is the maximum power gradient Δ_i^{max} describing by how much the production level can be increased within one time period. In order to avoid increased thermal stress of a power plant, the parameters θ_i^{up} and θ_i^{down} are used denoting the minimum running and the minimum down time, respectively. This means that once a plant starts to operate, it has to keep on running for at least θ_i^{up} time periods. Analogously, the plant must remain off for at least θ_i^{down} time periods, once it is turned off.

Considering the costs incurred by a power plant, we differentiate between fuel and variable costs, called γ_i^{fuel} and γ_i^{var}, respectively. Fuel costs refer to the costs per energy unit of the consumed energy, whereas variable costs are costs per energy unit produced during the planning horizon. Additionally, the start-up costs γ_i^{up} of a plant have to be taken into account, incurring a significant part of the total costs. Generally, these costs are expressed in dependence of the installed capacity p_i^{max}, i.e., considering the fuel costs for one hour of full capacity operation. Therefore, we assume that $\gamma_i^{up} = p_i^{max} \gamma_i^{fuel}$.

Energy Storages

The set $\mathcal{J}$ comprises all energy storages of the system. Being able to convert and store energy, a storage $j \in \mathcal{J}$ provides the possibility to supply energy when needed. Naturally, the amount of stored energy is bounded from below and above, represented by s_j^{min} and s_j^{max}, respectively. To prevent emptying the storage at the end of the planning horizon, we require the terminal storage level to equal $\bar{s}_{jT}$. An energy storage may contain more than one unit responsible for charging and discharging electrical power. Thus, for each storage $j \in \mathcal{J}$ we introduce the two sets $\mathcal{K}_j$ and $\mathcal{L}_j$ comprising all charging units and discharging units, respectively. Again, the produced power of these energy conversion units is bounded by $s_k^{in,min}$ and $s_k^{in,max}$ for a charging unit $k \in \mathcal{K}_j$ and analogously by $s_l^{out,min}$ and $s_l^{out,max}$ for a discharging unit $l \in \mathcal{L}_j$.

Starting up a storage unit consumes a certain amount of stored energy. This means, once it starts charging, α_j^{in} units of energy are used from the storage and once it starts discharging, α_j^{out} units are needed. Due to technical restrictions, the starting process of the storage may also produce additional costs. In that case, the corresponding costs per unit of energy, called $\gamma_j^{in,up}$ and $\gamma_j^{out,up}$, are considered in the objective function. Finally, certain types of discharging units consume additional energy to reproduce electricity from the stored energy. The parameter γ_j^{fuel} reflects the resulting costs in the objective function.

3.1.2 Variables and Efficiency Functions

In Table 3.3, a list of all variables which appear in the model is shown, starting with the continuous and ending with the binary variables. To each variable, a time index t is assigned, as all variables are needed for the description of each time step $t \in \mathcal{T}$.

For each power plant $i \in \mathcal{I}$, we introduce the continuous variable $p_{it} \in \mathbb{R}_+$ which represents the produced power. Additionally, the decision variable $z_{it} \in \{0,1\}$ is used to indicate whether the power plant is operating or not. Finally, we introduce the binary variables $z_{it}^{up} \in \{0,1\}$ and $z_{it}^{down} \in \{0,1\}$ modeling the switching process of the plant, respectively. In particular, $z_{it}^{up} = 1$ if and only if the plant is switched on in time period t and was not operating in period $t-1$. Analogously, z_{it}^{down} indicates if the plant is shut down in time t.

Table 3.3: Variables

p_{it}	produced power of plant $i \in \mathcal{I}$	[MW]
s_{jt}	storage level of storage $j \in \mathcal{J}$	[MWh]
s_{kt}^{in}	charging power of unit $k \in \mathcal{K}_j$ and $j \in \mathcal{J}$	[MW]
s_{lt}^{out}	discharging power of unit $l \in \mathcal{L}_j$ and $j \in \mathcal{J}$	[MW]
x_t	imported power	[MW]
c_{it}^{pow}	costs of plant $i \in \mathcal{I}$	[€]
c_{jt}^{stor}	costs of storage $j \in \mathcal{J}$	[€]
c_t^{imp}	import costs	[€]
z_{it}	state variable for the production of plant $i \in \mathcal{I}$	[1]
z_{it}^{up}	start-up variable of plant $i \in \mathcal{I}$	[1]
z_{it}^{down}	shut-down variable of plant $i \in \mathcal{I}$	[1]
z_{kt}^{in}	state variable of charging unit $k \in \mathcal{K}_j$ and $j \in \mathcal{J}$	[1]
z_{lt}^{out}	state variable of discharging unit $l \in \mathcal{L}_j$ and $j \in \mathcal{J}$	[1]
$z_{kt}^{in,up}$	start-up variable of charging unit $k \in \mathcal{K}_j$ and $j \in \mathcal{J}$	[1]
$z_{lt}^{out,up}$	start-up variable of discharging unit $l \in \mathcal{L}_j$ and $j \in \mathcal{J}$	[1]
y_{jt}^{in}	state variable for charging the storage $j \in \mathcal{J}$	[1]
y_{jt}^{out}	state variable for discharging the storage $j \in \mathcal{J}$	[1]

For each energy storage $j \in \mathcal{J}$, the variable $s_{jt} \in \mathbb{R}_+$ represents the current storage level. In order to describe the amount of charged power of a charging unit $k \in \mathcal{K}_j$ of a storage $j \in \mathcal{J}$, we use the variable $s_{kt}^{in} \in \mathbb{R}_+$. Likewise, the discharged power of discharging unit $l \in \mathcal{L}_j$ is described by $s_{lt}^{out} \in \mathbb{R}_+$. For the description of the operational state of a unit $k \in \mathcal{K}_j$, we introduce the decision variable $z_{kt}^{in} \in \{0,1\}$ and accordingly for a discharging unit $l \in \mathcal{L}_j$ the variable $z_{lt}^{out} \in \{0,1\}$. The start-up variables $z_{kt}^{in,up} \in \{0,1\}$ and $z_{lt}^{out,up} \in \{0,1\}$ indicate whether unit $k \in \mathcal{K}_j$ or $l \in \mathcal{L}_j$ is switched on in time period t, respectively. In order to describe whether any unit of storage j performs charging or discharging operations, the variables $y_{jt}^{in} \in \{0,1\}$ and $y_{jt}^{out} \in \{0,1\}$ are introduced. The amount of imported power procured in period t is represented by the variable $x_t \in \mathbb{R}_+$.

With regard to the objective function, the variable $c_{it}^{pow} \in \mathbb{R}_+$ models the costs arising from a power plant $i \in \mathcal{I}$. Likewise, $c_{jt}^{stor} \in \mathbb{R}_+$ represents the

Table 3.4: Efficiency functions

$\eta_i(p_{it})$	production efficiency of plant $i \in \mathcal{I}$
$\eta_j^{in}(s_{kt}^{in})$	charging efficiency of unit $k \in \mathcal{K}_j$ of storage $j \in \mathcal{J}$
$\eta_j^{out}(s_{lt}^{out})$	discharging efficiency of unit $l \in \mathcal{L}_j$ of storage $j \in \mathcal{J}$
$\eta_j^{ext}(s_{lt}^{out})$	discharging efficiency of unit $l \in \mathcal{L}_j$ of storage $j \in \mathcal{J}$ with respect to the external energy added

costs caused by storage $j \in \mathcal{J}$. Finally, the variable $c_t^{imp} \in \mathbb{R}$ describes the costs for the imported power.

The description of partial load efficiencies of facilities plays an important role within the model. As shown in Table 3.4, the given nonlinear functions appear in the description of the power plants as well as of energy storages. Basically, it expresses the ratio between the power input and the useful power output. As already mentioned, the efficiency of a power conversion machine depends on the current production level, in general showing a nonlinear behavior. As illustrated in Figure 3.1, the efficiency typically grows with increasing production.

For a power plant $i \in \mathcal{I}$, we represent the efficiency by the function $\eta_i(p_{it})$, depending on the produced power p_{it}. Here, the efficiency is associated with the ratio of the consumed power and the produced power p_{it}. Concerning a charging unit $k \in \mathcal{K}_j$ of storage $j \in \mathcal{J}$, the function $\eta_j^{in}(s_{kt}^{in})$ is introduced, reflecting the ratio between the consumed electric power s_{kt}^{in} and the converted power to be stored. Note that for each storage $j \in J$, we assume that all charging units $k \in \mathcal{K}_j$ are equal, i.e., having the same efficiency function η_j^{in}. Analogously, we introduce the function $\eta_j^{out}(s_{lt}^{out})$ expressing

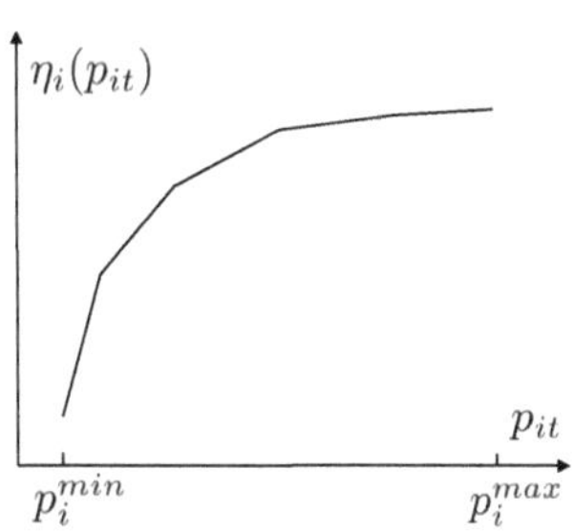

Figure 3.1: Efficiency of a power plant

the ratio of the power removed from the storage to the discharged power s_{lt}^{out}. Finally, the function $\eta_j^{ext}(s_{lt}^{out})$ describes the efficiency corresponding to the ratio of discharged power s_{lt}^{out} and power procured from outside. This power is considered in the objective function.

These nonlinear functions appear as univariate nonlinear terms in the model, which have to be approximated in an adequate way. In Section 3.1.5, we present a piecewise linear approximation of each nonlinear term, yielding a mixed-integer linear problem.

3.1.3 Constraints

In the following section the constraints describing the problem are modeled explicitly. We start with the major restriction, concerning the demand satisfaction. In each time step $t \in \mathcal{T}$ the demand δ_t has to be covered by the produced power p_{it} of the plants $i \in \mathcal{I}$, the imported power x_t, and the available wind power supply ω_t. The usage of the different types of energy storages $j \in \mathcal{J}$ is represented by the last two sums of the left hand side of the inequality. Thus, for all $t \in \mathcal{T}$, we obtain

$$\sum_{i \in \mathcal{I}} p_{it} + x_t + \omega_t + \sum_{j \in \mathcal{J}} \sum_{l \in \mathcal{L}_j} s_{lt}^{out} - \sum_{j \in \mathcal{J}} \sum_{k \in \mathcal{K}_j} s_{kt}^{in} \geq \delta_t. \tag{3.1}$$

Power Plants

For the description of a power plant $i \in \mathcal{I}$ the following constraints are required. In each time period $t \in \mathcal{T}$, the amount of produced power p_{it} is bounded by

$$z_{it} p_i^{min} \leq p_{it} \leq z_{it} p_i^{max}, \tag{3.2}$$

using the state variable z_{it}. Due to technical restrictions, the production level of a power plant cannot be increased arbitrarily within one time period. This is ensured by

$$p_{it} - p_{i,t-1} \leq \Delta_i^{max} z_{i,t-1} + p_i^{min}(1 - z_{i,t-1}), \tag{3.3}$$

for each time step $t \in \mathcal{T} \setminus \{1\}$. Here, the binary variables $z_{i,t-1}$ and $z_{i,t}$ are involved in order to suspend the power gradient restriction in case the plant is switched on in time step t.

In order to consider the minimum running time θ_i^{up} and minimum down time θ_i^{down} of a power plant $i \in \mathcal{I}$, the interconnection of the state variables

z_{it} corresponding to different time steps is described by

$$z_{it} - z_{i,t-1} \leq z_{in}, \quad \text{for } 2 \leq t < n \leq \min\{t + \theta_i^{up} - 1, T\}, \tag{3.4}$$

$$z_{i,t-1} - z_{it} \leq 1 - z_{in}, \quad \text{for } 2 \leq t < n \leq \min\{t + \theta_i^{down} - 1, T\}, \tag{3.5}$$

see for instance [MMM09, AC00]. Inequality (3.4) ensures the minimum running time restriction, forcing the right-hand side to equal one if the plant is operating in period t and was not running in $t-1$. Then, the power plant must run for the next θ_i^{up} time periods. Analogously, (3.5) models the minimum down time.

Starting a power plant causes additional costs γ_i^{up} which have to be considered in the objective function. Hence, the start-up variable z_{it}^{up} is needed. Additionally, shutting down a plant may produce costs, requiring the shut-down variable z_{it}^{down}. Both are described by the following constraints:

$$z_{it} - z_{i,t-1} - z_{it}^{up} + z_{it}^{down} = 0, \tag{3.6}$$

$$z_{it}^{up} + z_{it}^{down} \leq 1, \tag{3.7}$$

for $t \in \mathcal{T} \setminus \{1\}$, see for instance [LLM04, PW06, Mor07]. By (3.6) the decision variables z_{it}^{up} and z_{it}^{down} are connected to the state variable z_{it}. Finally, (3.7) ensures that a plant cannot be switched on and off at the same period t. Note that the constraints (3.6) and (3.7) represent a linearization of the constraints $z_{it}^{up} = z_{it}(1 - z_{i,t-1})$ and $z_{it}^{down} = z_{i,t-1}(1 - z_{it})$.

Energy Storages

In order to describe the properties of an energy storage $j \in \mathcal{J}$, we start with formulating the lower and upper bound on the storage level by

$$s_j^{min} \leq s_{jt} \leq s_j^{max}, \tag{3.8}$$

for all $t \in \mathcal{T}$. The next constraint bounds the terminal storage level from below:

$$s_{jT} \geq \bar{s}_{jT}. \tag{3.9}$$

It is needed to prevent the storage from being exhausted at the end of the planning horizon.

As already described in Section 3.1.1, a storage may consist of more than one unit responsible for charging or discharging. The lower and upper bounds on the corresponding charging variables s_{kt}^{in} for $k \in \mathcal{K}_j$ and discharging variables s_{lt}^{out} for $l \in \mathcal{L}_j$ is given by

$$
\begin{aligned}
s_k^{in,min} z_{kt}^{in} & \leq s_{kt}^{in} \leq & z_{kt}^{in} s_k^{in,max}, \quad (3.10)\\
s_l^{out,min} z_{lt}^{out} & \leq s_{lt}^{out} \leq & z_{lt}^{out} s_l^{out,max}, \quad (3.11)
\end{aligned}
$$

for all periods $t \in \mathcal{T}$.

The next constraint describes the conservation of energy of a storage $j \in \mathcal{J}$. It basically connects the storage level s_{jt} in time period $t \in \mathcal{T} \setminus \{1\}$ with the storage level $s_{j,t-1}$ in the previous time period $t-1$:

$$
\begin{aligned}
s_{jt} &= s_{j,t-1} + \frac{1}{\tau} \sum_{j \in \mathcal{J}} \left(\sum_{k \in \mathcal{K}_j} \eta_j^{in}(s_{kt}^{in}) s_{kt}^{in} - \sum_{l \in \mathcal{L}_j} \frac{1}{\eta_j^{out}(s_{lt}^{out})} s_{lt}^{out} \right) \\
&\quad - \sum_{j \in \mathcal{J}} \left(\sum_{k \in \mathcal{K}_j} \alpha_j^{in} z_{kt}^{in,up} + \sum_{l \in \mathcal{L}_j} \alpha_j^{out} z_{lt}^{out,up} \right). \quad (3.12)
\end{aligned}
$$

Here, $\eta_j^{in}(s_{kt}^{in})$ and $\eta_j^{out}(s_{lt}^{out})$ represent the efficiencies of the charging and discharging units. Usually, they are modeled via nonlinear, non-convex functions, depending on the charged power s_{kt}^{in} and the discharged power s_{lt}^{out}, respectively, see [HDS07]. The second line concerns the additional energy needed for starting a unit. For every start of a charging unit the energy α_j^{in} is consumed, which is taken from the stored energy. Therefore, for each unit $k \in \mathcal{K}_j$ and $t \in \mathcal{T} \setminus \{1\}$, the start-up variable $z_{kt}^{in,up}$ is needed, which is described as follows:

$$
\begin{aligned}
z_{kt}^{in} - z_{k,t-1}^{in} & \leq z_{kt}^{in,up}, \quad (3.13)\\
z_{kt}^{in,up} & \leq 1 - z_{k,t-1}^{in}, \quad (3.14)\\
z_{kt}^{in,up} & \leq z_{kt}^{in}. \quad (3.15)
\end{aligned}
$$

Analogously, α_j^{out} refers to the energy used for starting a discharging unit. Thus, for the description of the start-up variable $z_{lt}^{out,up}$ of unit $l \in \mathcal{L}_j$, we obtain:

$$
\begin{aligned}
z_{lt}^{out} - z_{l,t-1}^{out} & \leq z_{lt}^{out,up}, \quad (3.16)\\
z_{lt}^{out,up} & \leq 1 - z_{l,t-1}^{out}, \quad (3.17)\\
z_{lt}^{out,up} & \leq z_{lt}^{out}, \quad (3.18)
\end{aligned}
$$

for $t \in \mathcal{T} \setminus \{1\}$.

In order to connect the binary variables for charging and discharging to the state variables of storage j, we formulate:

$$z_{kt}^{in} \le y_{jt}^{in}, \qquad \text{for all } k \in \mathcal{K}_j, \tag{3.19}$$

$$z_{lt}^{out} \le y_{jt}^{out}, \qquad \text{for all } l \in \mathcal{L}_j, \tag{3.20}$$

for all $t \in \mathcal{T}$. Inequalities (3.19) force the state variable y_{jt}^{in} to one if at least one unit k of storage j is charging in t. Analogously, (3.20) refers to the discharging state y_{jt}^{out}.

In many cases, an energy storage j cannot be charged or discharged at the same time. With y_{jt}^{in} and y_{jt}^{out} describing the charging or discharging state of a storage, we obtain the following inequality:

$$y_{jt}^{in} + y_{jt}^{out} \le 1, \tag{3.21}$$

for all $t \in \mathcal{T}$.

Certain types of discharging units consume additional energy to reproduce electricity from the stored energy. By γ_j^{fuel} the resulting costs are considered in the objective function. In that case, the corresponding efficiency factor, called $\eta_j^{ext}(s_{lt}^{out})$ has to be taken into account in order to compute the consumed energy depending on the storage output s_{lt}^{out}.

Initial State

In order to be able to optimize the operation of the facilities, it is necessary to define the values of the variables at the beginning of the planning horizon, i.e., $t = 1$. This concerns only time dependent variables, as the remaining variables do not affect the optimization. For our model, we initialize the variables by

$$p_{i1} = \bar{p}_{i1}, \qquad \text{for all } i \in \mathcal{I}, \tag{3.22a}$$

$$z_{i1} = \bar{z}_{i1}, \qquad \text{for all } i \in \mathcal{I}, \tag{3.22b}$$

$$s_{j1} = \bar{s}_{j1}, \qquad \text{for all } j \in \mathcal{J}, \tag{3.22c}$$

$$z_{k1}^{in} = \bar{z}_{k1}^{in}, \qquad \text{for all } k \in \mathcal{K} \text{ and } j \in \mathcal{J}, \tag{3.22d}$$

$$z_{l1}^{out} = \bar{z}_{l1}^{out}, \qquad \text{for all } l \in \mathcal{L} \text{ and } j \in \mathcal{J}. \tag{3.22e}$$

Note that initializing the variables z_{i1} for all $i \in \mathcal{I}$ is not necessary, as this value directly follows from the initial production level $\bar{p}_{i1}$, assuming $p_i^{min} > 0$.

3.1.4 Objective Function

The major objective of this problem is the minimization of the total costs incurred by the power supply network, which basically consist of three parts: the costs of the plants, the costs of the storages, and finally the import costs. Concerning a power plant $i \in \mathcal{I}$ in time period $t \in \mathcal{T} \setminus \{1\}$, the costs c_{it}^{pow} are given by the sum of fuel costs, the variable costs, and the start-up cost:

$$c_{it}^{pow} = \gamma_i^{fuel} \frac{p_{it}}{\eta_i(p_{it})\,\tau} + \gamma_i^{var} \frac{p_{it}}{\tau} + \gamma_i^{up} y_{it}^{up}. \tag{3.23}$$

The fuel costs are computed with respect to the efficiency $\eta_i(p_{it})$, as they depend on the energy which is actually consumed by the plant. In contrast, the variable costs are expressed in dependence of the produced energy $\frac{p_{it}}{\tau}$ in period t. Note that the number of time steps per hour, denoted by τ, appears in the formulation.
The costs c_{jt}^{stor} of an energy storage $j \in \mathcal{J}$ in step $t \in \mathcal{T} \setminus \{1\}$ can be expressed as

$$c_{jt}^{stor} = \frac{\gamma_j^{fuel}}{\tau} \sum_{l \in \mathcal{L}_j} \frac{s_{lt}^{out}}{\eta_j^{ext}(s_{lt}^{out})} + \sum_{k \in \mathcal{K}_j} \gamma_j^{in,up} z_{kt}^{in,up} + \sum_{l \in \mathcal{L}_j} \gamma_j^{out,up} z_{lt}^{out,up}. \tag{3.24}$$

The first summand reflects the costs caused by the discharging of a unit $l \in \mathcal{L}_j$. These costs may appear if additional energy is necessary to reproduce electricity from the stored energy. The second and third term describe the start-up costs for charging and discharging, respectively. Here, constant but storage-dependent costs are assumed.

Finally, the costs for the imported power in period $t \in \mathcal{T} \setminus \{1\}$ are described by

$$c_t^{imp} = \gamma_t^{imp} \frac{x_t}{\tau}. \tag{3.25}$$

Thus, adding (3.23) for all power plants and (3.24) for all energy storages to (3.25) over the complete planning horizon, we obtain:

$$\min \sum_{t \in \mathcal{T} \setminus \{1\}} \left(\sum_{i \in \mathcal{I}} c_{it}^{pow} + \sum_{j \in \mathcal{J}} c_{jt}^{stor} + c_t^{imp} \right). \tag{3.26}$$

This completes the description of the MINLP formulation of the deterministic model.

3.1.5 Linearization of the Nonlinear Functions

In this section, we present the approximation of the nonlinear efficiency terms occurring within the description of power plants as well as in the description of energy storages. As the efficiency of a machine significantly depends on the current operation level, these functions are indispensable for a realistic problem description. Aiming at a mixed-integer linear formulation, we follow the approach of a piecewise linear approximation of the functions. First, we give a description of the nonlinear terms and their approximation and subsequently introduce the linearization method applied.

There are basically four nonlinear terms which have to be considered in the model. The first nonlinear term we are focusing on appears within the description of a power plant $i \in \mathcal{I}$. In the objective function, the term $\frac{p_{it}}{\eta_i(p_{it})}$ is used to compute the fuel costs c_{it}^{pow} of a power plant in a time step $t \in \mathcal{T} \setminus \{1\}$, see equation (3.23). Note that the nonlinear expression only depends on the produced power p_{it} and thus can be approximated by a univariate piecewise linear function denoted by

$$f_i(p_{it}) \approx \frac{p_{it}}{\eta_i(p_{it})},$$

where $f_i(p_{it})$ is defined on the domain $D_i = \{0\} \cup [p_i^{min}, p_i^{max}]$. The point $p_{it} = 0$ describes the off-position of the plant with $f_i(0) = 0$. Substituting the nonlinear term by $f_i(p_{it})$ in constraint (3.23), we obtain the linearized formulation

$$c_{it}^{pow} = \frac{\gamma_i^{fuel}}{\tau} f_i(p_{it}) + \frac{\gamma_i^{var}}{\tau} p^{it} + \gamma_i^{up} y_{it}^{up}, \tag{3.27}$$

for all $i \in \mathcal{I}$ and $t \in \mathcal{T} \setminus \{1\}$. Considering all power plants over the complete planning horizon, there are $|\mathcal{I}|(T-1)$ nonlinear terms which are approximated.

The second term $\frac{s_{lt}^{out}}{\eta_j^{ext}(s_{lt}^{out})}$ is used for the calculation of the costs c_{jt}^{stor}, resulting from the discharging units $l \in \mathcal{L}_j$ of a storage $j \in \mathcal{J}$ in time $t \in \mathcal{T} \setminus \{1\}$, see equation (3.24). Appearing in the objective function, the term represents the amount of additional power which is necessary to discharge s_{lt}^{out}. Using the approximation function

$$f_j^{ext}(s_{lt}^{out}) \approx \frac{s_{lt}^{out}}{\eta_j^{ext}(s_{lt}^{out})},$$

defined on $D_j^{ext} = \{0\} \cup [s_l^{out,min}, s_l^{out,max}]$, we obtain

$$c_{jt}^{stor} = \frac{\gamma_j^{fuel}}{\tau} \sum_{l \in \mathcal{L}_j} f_j^{ext}(s_{lt}^{out}) + \sum_{k \in \mathcal{K}_j} \gamma_j^{in,up} z_{kt}^{in,up} + \sum_{l \in \mathcal{L}_j} \gamma_j^{out,up} z_{lt}^{out,up}, \quad (3.28)$$

for all $j \in \mathcal{J}$ and $t \in \mathcal{T} \setminus \{1\}$. Observe that by considering all storages and time steps, $\sum_{j \in \mathcal{J}} |\mathcal{L}_j|(T-1)$ nonlinear terms are linearized.
Finally, the third and fourth nonlinear term occurs in the modeling of an energy storage, more precisely in the balance equation (3.12). Depending on the charged energy s_{kt}^{in} of a charging unit $k \in \mathcal{K}_j$, the term $s_{kt}^{in}\, \eta_j^{in}(s_{kt}^{in})$ computes the amount of power which is actually stored by unit k in time $t \in \mathcal{T} \setminus \{1\}$. Analogously, $\frac{s_{lt}^{out}}{\eta_j^{out}(s_{lt}^{out})}$ represents the amount of power taken from the storage in order to discharge s_{lt}^{out} by a discharging unit $l \in \mathcal{L}_j$. Both terms are approximated by the following piecewise linear functions:

$$\begin{aligned} f_j^{in}(s_{kt}^{in}) &\approx s_{kt}^{in}\, \eta_j^{in}(s_{kt}^{in}), \\ f_j^{out}(s_{lt}^{out}) &\approx \frac{s_{lt}^{out}}{\eta_j^{out}(s_{lt}^{out})}, \end{aligned}$$

which are defined on the domain $D_j^{in} = \{0\} \cup [s_k^{in,min}, s_k^{in,max}]$ and on $D_j^{out} = \{0\} \cup [s_l^{out,min}, s_l^{out,max}]$, respectively. Altogether, there are $\sum_{j \in \mathcal{J}}(|\mathcal{K}_j| + |\mathcal{L}_j|)(T-1)$ nonlinear terms, which have to be approximated. Replacing the terms by their corresponding approximation, the linearized balance equation yields

$$\begin{aligned} s_{jt} = s_{j,t-1} &+ \frac{1}{\tau} \sum_{j \in \mathcal{J}} \left(\sum_{k \in \mathcal{K}_j} f_j^{in}(s_{kt}^{in}) - \sum_{l \in \mathcal{L}_j} f_j^{out}(s_{lt}^{out}) \right) \\ &- \sum_{j \in \mathcal{J}} \left(\sum_{k \in \mathcal{K}_j} \alpha_j^{in} z_{kt}^{in,up} - \sum_{l \in \mathcal{L}_j} \alpha_j^{out} z_{lt}^{out,up} \right). \quad (3.29) \end{aligned}$$

A summary of all nonlinear terms and their corresponding linear approximations is given in Table 3.5.

Table 3.5: Approximation of nonlinear functions

piecewise linear approximation		nonlinear term
$f_i(p_{it})$	$\approx$	$\frac{p_{it}}{\eta_i(p_{it})}$
$f_j^{ext}(s_{lt}^{out})$	$\approx$	$\frac{s_{lt}^{out}}{\eta_j^{ext}(s_{lt}^{out})}$
$f_j^{in}(s_{kt}^{in})$	$\approx$	$s_{kt}^{in}\ \eta_j^{in}(s_{kt}^{in})$
$f_j^{out}(s_{lt}^{out})$	$\approx$	$\frac{s_{lt}^{out}}{\eta_j^{out}(s_{lt}^{out})}$

Modeling of Piecewise Linear Functions

Having introduced the approximation functions, it remains to describe their explicit mathematical formulation. There are various contributions which address the modeling of piecewise linear functions, most of them focusing on separable functions, see for instance [Dan63, Wil98, Pad00, MMM06, KdFN04]. For a recent overview on mixed-integer models for piecewise linear approximations we refer to [VAN09].

Basically, there are two well-known mixed-integer formulations describing the piecewise linear approximation: the *convex combination method* also called *lambda method*, see e.g. [Dan63, NW88], and the *incremental method* also known as *delta method*, see e.g. [Wil98]. The underlying idea of both methods is the approximation of the nonlinear function value by the convex combination of function values corresponding to the vertices of exactly one interval. In order to ensure that exactly one interval is chosen for the approximation, additional binary variables are introduced. A comparison of these approaches can be found in [Pad00, KdFN04].

Alternatively, there exists the possibility of modeling piecewise linear functions using so called Special Ordered Sets of Type 2 (SOS2), see [BT70, MMM06]. This method shows similarities to the convex combination method, except that no binary variables are used. Instead, the SOS2 condition is enforced algorithmically in the branch-and-bound phase. Recently, [VAN09] proposed an enhanced version of the convex combination method called *logarithmic method* requiring only a logarithmic number of binary variables and constraints instead of a linear number in the original method. For further information on piecewise linear approximation as well as for a discussion on the approximation error, we also refer to [GMMS09].

Table 3.6: Computational results applying different linearization methods

# Time steps	Incremental method	Logarithmic method	SOS method	Convex combination method
96	7.5 sec 120 nodes	14.1 sec 459 nodes	25.1 sec 1946 nodes	93.4 sec 3254 nodes
192	127.4 sec 1162 nodes	141.4 sec 5124 nodes	138.4 sec 6043 nodes	187.7 sec 1876 nodes
288	1000 sec 9562 nodes gap: 0.02 %	1000 sec 17450 nodes gap: 0.02 %	1000 sec 32254 nodes gap: 0.16 %	1000 sec 11548 nodes gap: 0.03 %

In order to select a suitable linearization method for the problem introduced in Section 3.1, the linearization approaches described above are compared, i.e., the *convex combination, incremental, SOS,* and *logarithmic method.* For the test runs the models are generated using ILOG Concert Technology 2.6 and for the solution ILOG CPLEX 11.1 is applied, see [CPL]. For the application of the SOS method, we use the internal SOS-formulation provided by ILOG CPLEX. The computations are done on a 2.4 GHz workstation with 1 GB of RAM.

We consider three different test instances varying in the number of time steps of the planning horizon, as indicated in the first column of Table 3.6. As one time steps corresponds to 15 minutes, the first test instance of 96 time steps covers the planning of one day. Analogously, the second test instance of 192 time steps corresponds to two days and the third one to three days. For detailed information on the test instance, we refer to Section 8.1.

In Table 3.6, the computational results consisting of the CPU time and the number of nodes of the branch-and-cut tree are listed. In case of the first two test instances, the problems are solved to optimality for all linearization techniques. For the largest test instance, the solution process was aborted after 1000 seconds. As a measure of the progress of the solution process, the gap computed by $\frac{UB-LB}{UB}$ is given, where LB denotes the best lower and UB the best upper bound found so far.

Regarding the running time for the first two instances as well as the gap for the last instance, the application of the incremental linearization method provides the best results. Additionally, the usage of this approximation method results in the lowest number of nodes of the corresponding branch-and-bound tree. These observations may follow from the strong linear relaxation of the incremental formulation, i.e., the extreme points of the sub-polyhedron defined by the linear relaxation are integral with respect to the components corresponding to binary variables, see [Pad00]. Due to the limited availability of data for the nonlinear efficiency functions, only up to ten grid points are used for the approximation. Thus, the impact resulting from the reduction of the number of binary variables in the logarithmic approach is not strong enough to provide the best solution times. The SOS approach yields good results for the first two test instances, but for the third instance the gap after 1000 seconds was even larger than applying the convex combination method, which provided the worst computational results for the first two instances. Based on these computational experiences, we select the incremental method for further computations and consequently restrict this section to the description of this approach.

In the following the classical formulation of the incremental method is adapted to the special structure of the nonlinear terms which appear in the model. To this end, the characteristics of the nonlinear terms are studied. First, all functions depend on one single variable each, and thus we restrict the description to the univariate case. Further on, all functions are defined on an interval $[a, b]$ united with the point $\{0\}$. Finally, the piecewise linear functions have to take the value zero if the associated state variable is zero, as this is true for all nonlinear functions. Based on these requirements, we obtain the following formulation of the piecewise linear approximation.

Let $h(x)$ be a nonlinear function of a continuous variable $x \in \mathbb{R}$, defined on $\{0\} \cup [a, b]$ with $0 < a < b$. Partitioning the interval $[a, b]$ by the grid points $a = a_0 < \ldots < a_K = b$, we obtain K subintervals $[a_{k-1}, a_k]$, with $k \in \mathcal{P} = \{1, \ldots, K\}$. For each subinterval $k \in \mathcal{P}$, we introduce a continuous variable δ_k and for each $k \in \mathcal{P} \setminus \{K\}$ a binary variable w_k. Note that for the last subinterval K no binary variable w_K is needed. In contrast to the textbook approach, a further binary variable z is included in order to ensure that $f(x)$ equals zero if $x = 0$. Then, the formulation of the piecewise linear approximation $f(x)$ of the nonlinear function $h(x)$ is given by:

$$x = az + \sum_{k=1}^{K} (a_k - a_{k-1})\delta_k, \tag{3.30a}$$

$$f(x) = h(a)z + \sum_{k=1}^{K} (h(a_k) - h(a_{k-1}))\delta_k, \tag{3.30b}$$

$$
\begin{aligned}
\delta_{k+1} \;\leq\; w_k \;\leq\; \delta_k, & \qquad \text{for all } k \in \mathcal{P} \setminus \{K\}, & (3.30c)\\
0 \;\leq\; \delta_k \;\leq\; 1, & \qquad \text{for all } k \in \mathcal{P}, & (3.30d)\\
az \;\leq\; x \;\leq\; bz, & & (3.30e)\\
\delta_1 \;\leq\; z, & & (3.30f)\\
w_k \;\in\; \{0,1\}, & \qquad \text{for all } k \in \mathcal{P} \setminus \{K\}, & (3.30g)\\
z \;\in\; \{0,1\}. & & (3.30h)
\end{aligned}
$$

The first equation (3.30a) describes the variable x in dependence of δ_k. Then, $f(x)$ can be formulated as a piecewise linear function, described in (3.30b). Inequalities (3.30c) are called the *filling condition*, see [Wil98], ensuring that if an interval k is chosen for the approximation, all intervals l with $l < k$ are used, i.e. $\delta_l = 1$. Conditions (3.30e) connect x and z by requiring that $x = 0$ if and only if $z = 0$. Finally, condition (3.30f) ensures that if $z = 0$, $\delta_1 = 0$ and thus all $\delta_i = 0$. Consequently, the requirement that $f(x) = 0$ if $x = 0$ is satisfied.

Remember that in the nonlinear model, for each continuous variable x of a nonlinear term $h(x)$, there already exists a binary variable z describing its status. For instance, the binary variable z_{it} of a power plant is connected to the production variable p_{it}. Furthermore, the bounds (3.30e) are also part of the original model. Consequently, for each piecewise linear approximation, additionally K continuous and $K-1$ binary variables are introduced.

Finally, we remark that due to limited availability of data, the efficiency functions are not given as an explicit function. Instead, the efficiency of the facilities is known only in selected operating points, providing the basis for the piecewise linear approximation.

3.1.6 The *D-OPGen* Model

Summarizing, in Section 3.1.3 a mixed-integer nonlinear model is formulated consisting of constraints (3.1) to (3.25) with objective function (3.26). Based on the piecewise linear approximation of the nonlinear terms presented in Section 3.1.5, now we can replace the affected constraints by their linearized version. In particular, we substitute

constraints (3.23) describing the costs of a power plant	by (3.27),
constraints (3.24) describing the costs of an energy storage	by (3.28),
and the balance equations (3.12) of an energy storage	by (3.29).

Remember that for each approximation of a nonlinear term by a piecewise linear function the system of constraints (3.30) has to be added to the problem. Altogether, we obtain a MILP formulation, which we denote by *D-OPGen* (**D**eterministic **O**ptimization of **P**ower **Gen**eration). This completes the modeling of the deterministic problem.

3.2 Stochastic Model

In this section, we present an enhanced version of the *D-OPGen* problem, including uncertainty with regard to some parameters. Among others mainly load profiles, electricity prices, and the power supply by regenerative energy are affected. In this thesis, we focus on the optimization under uncertain wind power generation, resulting from its meteorological dependence and uncertain prices for electricity. Regarding the fluctuations of the consumers' demand, we assume the corresponding load to be deterministic, as a reliable forecast of the load of one day or even one week can be given by schedulers.

3.2.1 Basic Concepts in Stochastic Programming

The following introduction to the basic concepts in stochastic programming and the notation used in this thesis are based on [RS01], [BL97] and [KM05]. In order to include uncertainty in the modeling of the *D-OPGen* problem, a probabilistic description is used. Therefore, we assume that the uncertain information is given by a discrete time stochastic process $\boldsymbol{\xi}$ defined on a probability space $(\Omega, \mathcal{F}, \mathcal{P})$ with

$$\boldsymbol{\xi} = \{ \, \xi_t := (\omega_t, \gamma_t^{imp}) \, \}_{t \in \mathcal{T}}.$$

Here, the random variable ξ_t describes the uncertain data in time step $t \in \mathcal{T} = \{1, \ldots, T\}$ taking values in $\mathbb{R}^2$. In our case, ω_t represents the wind power available in period t and γ_t^{imp} the price for electricity in t.

At the beginning of the planning horizon only the data for time step $t = 1$ is known, which means that ξ_1 is deterministic. For data of future periods only the probability distribution is given. Nevertheless, decisions on how to operate the facilities have to be made without complete knowledge of the wind power production or electricity prices during the planning horizon. Transferred to time period t this means that in order to make a decision in t only the realizations of the stochastic data up to this period can be taken into account. Thus, we assume the decisions x_t to be *nonanticipative*, i.e., to depend only on $\xi^t = (\xi_1, \ldots, \xi_t)$.

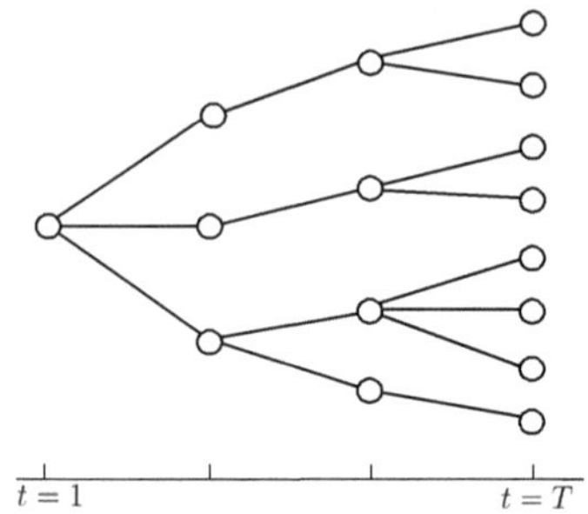

Figure 3.2: Four-stage scenario tree with eight leaves

Following a common approach in multistage stochastic programming to make the problem computationally manageable, we additionally assume that $(\Omega, \mathcal{F}, \mathcal{P})$ has the following properties: Let Ω be finite, which means $\Omega = \{\omega_s\}_{s \in \mathcal{S}}$ with $\mathcal{S} = \{1, \ldots, S\}$, $\mathcal{F}$ be the power set of Ω and finally let $\mathcal{P}(\{\omega\}) = p_s$ with $s \in \mathcal{S}$. By $\{\mathcal{F}_t\}_{t \in \mathcal{T}}$ we denote the filtration induced by $\boldsymbol{\xi}$, where $\mathcal{F}_t \subseteq \mathcal{F}$ is the σ-algebra generated by ξ^t, i.e. the information observable until period $t \in \mathcal{T}$. As ξ_1 is deterministic, we have $\mathcal{F}_1 = \{\emptyset, \Omega\}$ and assuming to have full information at the end of the planning horizon we require $\mathcal{F}_T = \mathcal{F}$. By ξ_t^s we denote the value of the data scenario s at time t with $s \in \mathcal{S}$ and $t \in \mathcal{T}$. Here, a scenario $\xi^s = (\xi_1^s, \ldots, \xi_T^s)$ corresponds to a realization of the process over the complete planning horizon $\mathcal{T}$. Regarding the fan formed by the individual scenarios ξ^s, the structure results in a tree by merging all scenarios up to period t which coincide until t, i.e., they are combined to a path. Hence, this tree is called *scenario tree.*

A scenario tree is denoted by $\Gamma = (\mathcal{N}, \mathcal{A})$ and is based on a finite number of nodes $\mathcal{N}$. The set $\mathcal{A}$ contains all arcs of the tree. In detail a scenario tree is given by a rooted tree with T layers, where each layer corresponds to a period t of the program. The root node $n = 1$ corresponds to time period $t = 1$ and $t(n)$ denotes the time stage of node n. As Γ is a tree, each node $n \in \mathcal{N}$ has a unique predecessor $p(n)$. Generalizing, the k-th predecessor of a node is denoted by $p^k(n)$. The set $\mathcal{N}_t$ contains all nodes of period t. Consequently, $\mathcal{N}_T$ consists of all leaf nodes of Γ, which means that the corresponding nodes do not have a successor. An example of a scenario tree with four layers and eight leaves is illustrated in Figure 3.2. Each path from the root node to a leaf node is associated with exactly one scenario, which represents a realization of the uncertain parameters over the whole planning horizon, i.e., if there are S leaf nodes in Γ, there are

Table 3.7: Notation for the stochastic problem

$\Gamma = (\mathcal{N}, \mathcal{A})$	rooted tree with nodes $\mathcal{N}$, arcs $\mathcal{A}$ and root node $n = 1$
$\mathcal{N}_t$	set of nodes of time stage t
$\mathcal{S}$	index set of scenarios
$t(n)$	time stage of node n
$p(n)$	predecessor of node n
$p^k(n)$	k-th predecessor of node n
$path(n)$	set of nodes of path $(1, n)$
π_n	probability of node n

S corresponding scenarios s with $s \in \mathcal{S} = \{1, \ldots, S\}$. Additionally, we denote the set of nodes corresponding to a path from the root node to n by $path(n)$. By π_n, we refer to the probability of a scenario to pass a node $n \in \mathcal{N}$. Consequently, the probability of the root equals 1, i.e., $\pi_1 = 1$. An overview of the notation introduced for the stochastic problem is given in Table 3.7.

Using this scenario tree notation, the following block-structured MIP describes a typical multistage stochastic mixed-integer problem:

$$
(SMIP) \qquad \begin{aligned}
\min \; & \sum_{n \in \mathcal{N}} \pi_n c_n^\top x_n \\
\text{s.t.} \; & \begin{array}{rcll}
W_1 x_1 & = & b_1 & \\
T_n x_{p(n)} + W_n x_n & = & b_n & \text{for all } n \in \mathcal{N} \setminus \{1\} \\
x_n & \in & X_n & \text{for all } n \in \mathcal{N}
\end{array}
\end{aligned}
$$

where x_n denotes the decision variables of node n and T_n and W_n are matrices of corresponding size as well as the vector c_n. The set X_n represents the restrictions requiring some or all of the variables of node n to be integer. For further information regarding stochastic programming, we refer to [LL93] and [KM05].

3.2.2 The *S-OPGen* Model

With the notation described above, we can formulate a multistage stochastic model for our problem. The variables are no longer only assigned to a time step $t \in \mathcal{T}$, but also depend on the scenarios of the stochastic process $\boldsymbol{\xi}$ represented by the scenario tree $\Gamma = (\mathcal{N}, \mathcal{A})$. If we use the nonanticipativity condition requiring that a decision in period t may only depend on realizations of the stochastic data up to t, the variables can carry the node index $n \in \mathcal{N}$, instead of index $t \in \mathcal{T}$ and $s \in \mathcal{S}$. Remember that a path $(1, n)$ in the scenario tree combines all those scenarios which share the same history.

Clearly, these variables have to satisfy the constraints introduced in Section 3.1.3. Expressing them based on the scenario tree yields the following stochastic formulation: Concerning the demand condition (3.1), we obtain

$$\sum_{i \in \mathcal{I}} p_{in} + x_n + \omega_n + \sum_{j \in \mathcal{J}} \sum_{l \in \mathcal{L}_j} s_{ln}^{out} - \sum_{j \in \mathcal{J}} \sum_{k \in \mathcal{K}_j} s_{kn}^{in} \geq \delta_{t(n)},$$

for each $n \in \mathcal{N} \setminus \{1\}$. Note that the parameter ω_n describing the wind power carries the node index n as all variables. By representing the wind power realization of the stochastic process corresponding to node n, it varies in accordance with the scenarios associated with node n. In contrast, the demand $\delta_{t(n)}$ has the same value for all nodes n within the same time stage, assuming δ to be deterministic.

This procedure can be applied to all constraints which do not correspond to more than one time stage, i.e., involving only variables which are associated with exactly one time stage. In particular this concerns the constraints describing the

lower and upper bound on the production variable of a plant	(3.2),
connection of the start-up and shut-down variables of a plant	(3.7),
lower and upper bound on the storage level of a storage	(3.8),
final storage level of a storage	(3.9),
lower and upper bound on the power of a charging unit	(3.10),
lower and upper bound on the power of a discharging unit	(3.11),
connection of the storage start-up and charging state	(3.15),
connection of the storage shut-down and discharging state	(3.18),
connection of charging units and charging state of a storage	(3.19),
connection of discharging units and discharging state of a storage	(3.20),
connection of charging and discharging storage state	(3.21),
and the initialization of the time step $t = 1$	(3.22).

For dynamic constraints which contain variables associated with two consecutive time steps, additionally the former index $t-1$ has to be replaced by the index $p(n)$, denoting the predecessor of node n. This procedure affects the following constraints describing the

upper bound on the power gradient of a plant (3.3),
connection of the start-up, shut-down, and the state of a plant (3.6),
lower bound on the start-up variable of charging unit (3.13),
upper bound on the start-up variable of a charging unit (3.14),
lower bound on the start-up variable of a discharging unit (3.16),
upper bound on the start-up variable of a discharging unit (3.17),
and the storage balance restriction (3.29).

Now, we focus on those constraints connecting more than two consecutive time steps. In the deterministic problem the minimum running time constraints (3.4) and the minimum down time constraints (3.5) show this characteristic. The derived stochastic formulation for the minimum running time conditions of a power plant $i \in \mathcal{I}$ yields

$$z_{in} - z_{i,p(n)} \leq z_{ik},$$

for all node pairs $(n, k) \in \mathcal{N} \times \mathcal{N}$ with $n \in path(k)$ and which satisfy $2 \leq t(n) < t(k) \leq \min\{t(n) + \theta_i^{up} - 1, T\}$. Remember that θ_i^{up} denotes the minimum running time of plant i. Corresponding to the minimum down time restriction, the constraint is reformulated by

$$z_{i,p(n)} - z_{in} \leq 1 - z_{ik},$$

for all node pairs $(n, k) \in \mathcal{N} \times \mathcal{N}$ with $n \in path(k)$ and which satisfy $2 \leq t(n) < t(k) \leq \min\{t(n) + \theta_i^{down} - 1, T\}$. Parameter θ_i^{down} represents the minimum down time, respectively.

Concerning the objective function, it is reasonable to minimize the cost arising in time stage $t = 1$ together with the expected costs of the time stages $t = 2$ to T. Each of the summands (3.25), (3.27), and (3.28) of the objective function (3.26), associated with the deterministic problem, are adapted to the scenario tree formulation described above. Hence, the objective function of the stochastic problem is expressed by

$$\min \sum_{n \in \mathcal{N}} \pi_n \left(\sum_{i \in \mathcal{I}} c_{in}^{pow} + \sum_{j \in \mathcal{J}} c_{jn}^{stor} + c_n^{imp} \right). \tag{3.31}$$

Note that as for the deterministic problem, the nonlinearities of the stochastic problem are approximated by the piecewise linear functions introduced in Section 3.1.5. For the modeling of the approximation functions also the incremental method is chosen because the good computational results are

Table 3.8: Computational results corresponding to the stochastic problem applying different linearization methods

# Time steps	Incremental method	Logarithmic method	SOS method	Convex combination method
12	1.4 sec 34 nodes	1.4 sec 43 nodes	1.5 sec 170 nodes	2.5 sec 48 nodes
24	86.5 sec 771 nodes	136.7 sec 2995 nodes	153.8 sec 9907 nodes	150.7 sec 2571 nodes

confirmed by the results obtained for two selected stochastic test instances, see Table 3.8. For more information concerning the test runs, we refer to Section 3.1.5 and for details of the test instances see Section 8.1. This completes the description of the stochastic problem, which we denote by *S-OPGen* (**S**tochastic **O**ptimization of **P**ower **Gen**eration).

In summary, using the scenario tree formulation, the problem can be expressed as a large scale deterministic MILP, which can be solved using standard MILP-solvers. As this approach might be computationally very expensive, a solution method is developed in Chapter 6, making use of the special structure of the multistage stochastic mixed-integer problem.

Chapter 4

Polyhedral Study of Stochastic Switching Polytopes

In this chapter, we investigate the solution set of the minimum runtime and downtime conditions of a power plant, introduced in Section 3.1.3. When a plant is switched on, these restrictions ensure that the plant keeps running for a certain number of time steps and when it is turned off, it must remain off for a certain number of time steps, too. In our problem these restrictions have to be considered for the coal power plant in order to avoid increased thermal stress. In this chapter, we study the underlying $0/1$ polytope for the stochastic formulation where uncertainty is modeled by a set of scenarios, as described in Section 3.2. Thus, we will call it *stochastic switching polytope*. For our studies, we used the software packages PORTA and POLYMAKE, see [CL] and [GJ00], in order to obtain a complete linear description of small instances.

This chapter is organized as follows. We start with a mathematical formulation of the stochastic switching polytope. Afterwards we give a literature survey, which concentrates on contributions addressing the corresponding deterministic formulation. The following section focuses on the investigation of the facial structure of the underlying polytope, and we present a linear description of the polytope. Finally, we provide an efficient separation algorithm, which detects the maximally violated inequality.

4.1 Mathematical Formulation

Let Γ be a scenario tree, based on the set of nodes $\mathcal{N} = \{1, \ldots, N\}$, i.e., $\Gamma = (\mathcal{N}, \mathcal{E})$. Then, a point in the stochastic switching polytope is determined by the values assigned to the binary variables x_n for $n \in \mathcal{N}$ and to the binary variables x_n^{up} and x_n^{down} for $n \in \mathcal{N} \setminus \{1\}$. The variable x_n represents the state of a plant in node n, i.e., $x_n = 1$ if the plant is operating and $x_n = 0$ else. The variable x_n^{up} indicates whether a plant is switched on in node n and thus, was not running in the predecessor node $p(n)$. Analogously, x_n^{down} states if the plant is switched off in n. For a better overview, we define

$$x := (x_1, \dots, x_N, x_2^{up}, \dots, x_N^{up}, x_2^{down}, \dots, x_N^{down})^\top, \tag{4.1}$$

which comprises all variables of the studied substructure. For a shorter description of the minimum run- and downtime, in this chapter we use the following notation: By $L \in \mathbb{N}$ we denote the minimum runtime of a plant and by $l \in \mathbb{N}$ the minimum downtime, respectively, formerly θ_i^{up} and θ_i^{down} in Section 3.1.1. Based on these parameters, we define the following two sets:

$$\begin{aligned}
\mathcal{L}_L^{up} := \; & \{(n,k) \in \mathcal{N} \times \mathcal{N} \mid \; n \in path(k), \\
& \qquad 2 \leq t(n) < t(k) \leq \min\{t(n) + L - 1, T\}\}, \\
\mathcal{L}_l^{down} := \; & \{(n,k) \in \mathcal{N} \times \mathcal{N} \mid \; n \in path(k) \\
& \qquad 2 \leq t(n) < t(k) \leq \min\{t(n) + l - 1, T\}\}.
\end{aligned}$$

Remember that the set $path(k)$ contains all nodes of the path $(1, \dots, k)$ on the scenario tree Γ and $t(n)$ indicates the time stage of n. In the following we restrict ourselves to the condition that $T - L \geq 1$ and $T - l \geq 1$ which is reasonable for the description of our problem.

Now, we can formulate the following constraints in order to model the minimum runtime and downtime restrictions, as already described in Section 3.2.2:

$$x_n - x_{p(n)} \leq x_k, \quad \text{for} \quad (n,k) \in \mathcal{L}_L^{up}, \tag{4.2}$$

$$x_{p(n)} - x_n \leq 1 - x_k, \quad \text{for} \quad (n,k) \in \mathcal{L}_l^{down}, \tag{4.3}$$

$$x_n - x_{p(n)} - x_n^{up} + x_n^{down} = 0, \quad \text{for} \quad n \in \mathcal{N} \setminus \{1\}, \tag{4.4}$$

$$x_n^{up} + x_n^{down} \leq 1, \quad \text{for} \quad n \in \mathcal{N} \setminus \{1\}, \tag{4.5}$$

where $p(n)$ denotes the predecessor of node n. The inequality (4.2) ensures the minimum runtime restriction. It forces the right hand side to equal one, if $x_n = 1$ and $x_{p(n)} = 0$, i.e., regarding node n, the plant is operating and was not running one time step before. In this case, by definition

of the set $\mathcal{L}_L^{up}$, the variables x_n have to equal one in all descendant nodes of n within the next L time steps, which means that the power plant is operating regarding these nodes. Inequality (4.3) describes the minimum downtime, respectively. The x_n^{up} and x_n^{down} variables are connected to the state variable x_n by equation (4.4). Finally, inequality (4.5) ensures that a plant cannot be switched on and off at the same node n.
The polytope defined by the convex hull of the feasible points of constraints (4.2) to (4.5) is denoted by $P_{\Gamma,L,l}$, which means

$$P_{\Gamma,L,l} = \operatorname{conv}\{x \in \{0,1\}^{3N-2} \mid x \text{ satisfies conditions (4.2) to (4.5)}\}.$$

4.2 Literature Overview

The minimum runtime and downtime restrictions of power plants play an important role in energy production problems, particularly when thermal power plants are involved, see for example [GNRS00, GKKN$^+$02, HNNS06]. But there are also further applications, for instance in gas network optimization, [Mor07], where due to technical restrictions of the compressors, the minimum up and down time conditions have to be taken into account.

Concerning the polyhedral structure, in [LLM04] the authors investigate the 0/1 polytopes, called *min-up/min-down polytopes*, which are described by inequalities (4.2) and (4.3) for the deterministic formulation. This corresponds to the scenario tree formulation, where Γ consists of only one scenario. Consequently, in the deterministic case the number of time steps $T \in \mathbb{N}$ of the planning horizon equals the number of nodes N. More precisely, the authors analyze the facial structure of the convex hull of the solution set and provide a complete linear description of the polytope. Additionally, a linear time separation algorithm is presented.

If switching costs are considered, the additional binary variables x^{up} and x^{down} can be used in order to model the start up or shut down of a machine, as described by (4.4) and (4.5). For the deterministic case, we denote the convex hull of all feasible solutions of the system of inequalities (4.2) to (4.5) by $P_{T,L,l}$, where T is the number of time steps. This polytope was independently investigated by [LLM04] and [Mor07], yielding a complete linear description of $P_{T,L,l}$. As these facet-defining inequalities provide the basis for our investigation of the stochastic switching polytope $P_{\Gamma,L,l}$, subsequently we specify the main results of these papers. Therefore, we basically follow the notation of [Mor07] which we adjust to our scenario tree formulation.

As the deterministic case is considered, there is only one scenario in the scenario tree formulation. Thus, here the number of times steps T equals the number of nodes N of the associated scenario tree Γ. Using the variables x_t, $t \in \{1,\ldots,T\}$ and x_t^{up}, x_t^{down}, $t \in \{2,\ldots,T\}$ introduced above, the deterministic switching polytope is defined by

$$P_{T,L,l} = \text{conv}\ \{x \in \{0,1\}^{3T-2} | \ x \text{ satisfies (4.2) to (4.5)}\}, \tag{4.6}$$

where $x = (x_1,\ldots,x_T,x_2^{up},\ldots,x_T^{up},x_2^{down},\ldots,x_T^{down})^\top$ and L is the minimum runtime and l the minimum downtime. For this polytope, the authors proof that the following inequalities are facet-defining:

$$x_t^{up}, x_t^{down} \geq 0, \quad \text{for} \quad t = 2,\ldots,T, \tag{4.7}$$

$$-x_T + \sum_{k=i}^{T} x_k^{up} - \sum_{k=i+L}^{T} x_k^{down} \leq 0, \quad \text{for} \quad i = 2,\ldots,T-L+1, \tag{4.8}$$

$$x_T - \sum_{k=i+l}^{T} x_k^{up} + \sum_{k=i}^{T} x_k^{down} \leq 1, \quad \text{for} \quad i = 2,\ldots,T-l+1. \tag{4.9}$$

In fact, they show that these inequalities together with equations (4.4) completely describe $P_{T,L,l}$. In [Mor07], this result is shown by proving that the resulting system of linear inequalities is totally dual integral. For more details concerning this proof, we also refer to [Mar05], who investigated the deterministic switching polytope $P_{T,L,l}$ within his diploma thesis. In contrast [LLM04] proved that each point in the polytope can be written as a convex combination of integral elements of $P_{T,L,l}$.

4.3 Polyhedral Investigations

In this section we focus on the investigation of the facial structure of $P_{\Gamma,L,l}$, where Γ represents the corresponding scenario tree. A scenario $s \in \mathcal{S}$ of the scenario tree Γ can be represented by a path $(v_1,\ldots,v_T)$ from the root node v_1 to the corresponding leaf node v_T. According to the definition of x in (4.1), we define

$$x^s := (x_{v_1},\ldots,x_{v_T},x_{v_2}^{up},\ldots,x_{v_T}^{up},x_{v_2}^{down},\ldots,x_{v_T}^{down})^\top$$

in order to refer to those variables which correspond to the scenario s.

Thereon, the stochastic switching polytope $P_{\Gamma_s,L,l}$ associated with one scenario is defined by

$$P_{\Gamma_s,L,l} = \text{conv}\{x^s \in \{0,1\}^{3T-2} \mid x^s \text{ satisfies conditions (4.2) to (4.5)}\},$$

where Γ_s is the scenario tree induced by the nodes $\{v_1, \ldots, v_T\}$, which are associated with scenario s. Particularly, here the constraints (4.2) to (4.5) refer to the restricted index set $\mathcal{N}_s := \{v_1, \ldots, v_T\}$. Note that this structure corresponds to the deterministic switching polytope defined in (4.6). Indeed, the deterministic switching polytope it is a special case of the stochastic switching polytope, where the scenario tree consists of a single path.

Now, the facet-defining inequalities (4.7) to (4.9) are adapted to the stochastic structure by

$$x_n^{up} \geq 0, \qquad x_n^{down} \geq 0, \tag{4.10}$$

for all $n \in \mathcal{N} \setminus \{1\}$ and by

$$-x_{v_T} + \sum_{k=i}^{T} x_{v_k}^{up} - \sum_{k=i+L}^{T} x_{v_k}^{down} \leq 0, \quad \text{for } i = 2, \ldots, T-L+1, \tag{4.11}$$

$$x_{v_T} - \sum_{k=i+l}^{T} x_{v_k}^{up} + \sum_{k=i}^{T} x_{v_k}^{down} \leq 1, \quad \text{for } i = 2, \ldots, T-l+1. \tag{4.12}$$

for all $s \in \mathcal{S}$ with corresponding path $(v_1, \ldots, v_T)$.

For these $2N-2+|\mathcal{S}|(2T-L-l)$ inequalities, we show that they also define facets of the stochastic polytope $P_{\Gamma,L,l}$. For the proof, the following points are needed, satisfying all constraints of $P_{\Gamma,L,l}$. First we consider the trivial points in $\mathbb{R}^{3N-2}$

$$e^{on} := (1, \ldots, 1, 0, \ldots, 0, 0, \ldots, 0)^\top,$$

which corresponds to a power plant which is operating in all time steps and

$$e^{off} := (0, \ldots, 0, 0, \ldots, 0, 0, \ldots, 0)^\top,$$

where the plant is turned off over the whole planning horizon.

Further on, we define the points a_n^{down} for all $n \in \mathcal{N} \setminus \{1\}$ which are associated with the decision that the power plant is switched off in node n, i.e., $x_n^{down} = 1$. We also require that the power plant is turned off in all nodes $k \in desc(n)$ and is operating in all other nodes. Recall that the set $desc(n)$ contains all successor nodes of n, i.e., all nodes of the subtree rooted in n. Thus, a_n^{down} is a feasible point in $P_{\Gamma,L,l}$. Analogously, we define the point a_n^{up} for which the plant is switched on in node n, i.e., $x_n^{up} = 1$, and is operating in all nodes $k \in desc(n)$. Additionally, the plant is not running in all other nodes. Figure 4.1 graphically shows an example of the points a_n^{up}

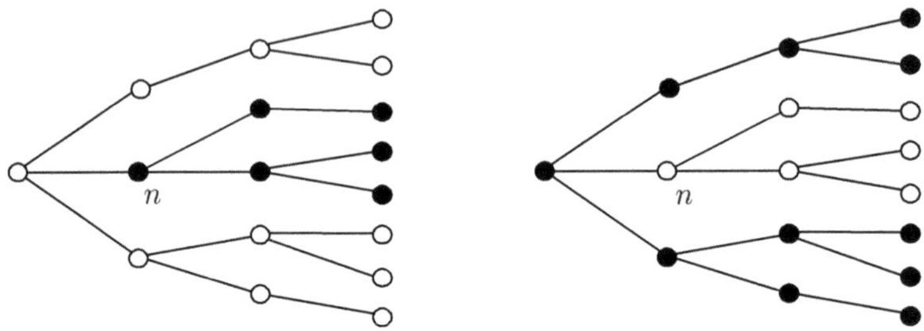

Figure 4.1: Scenario tree representing the points a_n^{up} and a_n^{down}

and a_n^{down}, where black nodes refer to an operating power plant and white nodes to a non-operating one.

We start by proving that the dimension of the polytope $P_{\Gamma,L,l}$ is equal to $2N-1$.

Lemma 4.1 *Let* Γ *be a scenario tree with* $N \in \mathbb{N}$ *nodes and let* $l, L \in \mathbb{N}$. *Then* $\dim(P_{\Gamma,L,l}) = 2N-1$.

Proof. Due to the number of variables, $\dim(P_{\Gamma,L,l}) \leq 3N-2$. Additionally, we know that there are $N-1$ equations of type (4.4), which are clearly linearly independent. This leads to $\dim(P_{\Gamma,L,l}) \leq 2N-1$. Now, we can specify the following $2N$ affinely independent points: e^{on}, e^{off} and a_n^{down}, a_n^{up} for $n \in \mathcal{N} \setminus \{1\}$. In order to show that these points are affinely independent, we neglect to point e^{off} which contains only zero entries and show linear independence. Therefore, we consider the matrix which consists of the selected points $(e^{on}, a_2^{up}, \ldots a_N^{up}, a_2^{down}, \ldots, a_N^{down})$ yielding

$$\begin{pmatrix} 1 & A^{up} & A^{down} \\ 0 & I_{N-1} & 0 \\ 0 & 0 & I_{N-1} \end{pmatrix},$$

where $A^{up} \in \mathbb{R}^{N\times(N-1)}$ and $A^{down} \in \mathbb{R}^{N\times(N-1)}$. Neglecting the first $N-1$ rows, we obtain an upper diagonal matrix with ones on the main diagonal which implies that the original matrix has full column rank. Thus, the selected points are linearly independent and $\dim(P_{\Gamma,L,l}) = 2N-1$. □

Based on the dimension of $P_{\Gamma,L,l}$, we know that a facet has dimension $2N-2$. Thus, the basic idea of the following proof is to show that for each inequality there are $2N-1$ affinely independent points of $P_{\Gamma,L,l}$ for which the inequality is tight.

Lemma 4.2 *For each node $n \in \mathcal{N} \setminus \{1\}$, inequalities (4.10) define facets of the polytope $P_{\Gamma,L,l}$. Additionally, for each scenario $s \in \mathcal{S}$ with corresponding path $(v_1, \dots, v_T)$, inequalities (4.11) and (4.12) also define facets of the polytope $P_{\Gamma,L,l}$.*

Proof. Since the inequalities are valid for the corresponding deterministic switching polytope $P_{\Gamma_s,L,l}$, they are also valid for $P_{\Gamma,L,l}$. This is true as the system of constraints describing $P_{\Gamma_s,L,l}$ is a subsystem of the constraints describing $P_{\Gamma,L,l}$.

For each inequality, we choose $2N-1$ affinely independent points, which satisfy the corresponding inequality by equality.

Inequalities (4.10)

At first, we concentrate on the nonnegativity constraint $x_n^{up} \geq 0$ for a fixed $n \in \mathcal{N} \setminus \{1\}$. For this constraint we choose the $N-1$ points a_i^{down} for $i \in \mathcal{N} \setminus \{1\}$ and the $N-2$ points a_i^{up} for $i \in \mathcal{N} \setminus \{1, i\}$. Together with the points e^{on} and e^{off} we obtain $2N-1$ points.

In analogy to the proof of Lemma 4.1, we prove affine independence by discarding the point e^{off} and showing linear independence of the remaining points. Thus, we consider the matrix corresponding to the points $(e^{on}, a_2^{up}, \dots, a_{n-1}^{up}, a_{n+1}^{up}, \dots a_N^{up}, a_2^{down}, \dots, a_N^{down})$ and obtain

$$\begin{array}{r} \\ \\ n \rightarrow \\ \\ \\ \end{array} \begin{pmatrix} \mathbf{1} & \multicolumn{2}{c}{A^{up}} & A^{down} \\ \mathbf{0} & \boldsymbol{I}_{n-2} & \mathbf{0} & \mathbf{0} \\ 0 & 0 \cdots 0 & 0 \cdots 0 & 0 \cdots 0 \\ \mathbf{0} & \mathbf{0} & \boldsymbol{I}_{N-n} & \mathbf{0} \\ \mathbf{0} & \mathbf{0} & \mathbf{0} & \boldsymbol{I}_{N-1} \end{pmatrix},$$

where $A^{up} \in \mathbb{R}^{N \times N-2}$ and $A^{down} \in \mathbb{R}^{N \times N-1}$. Again, we obtain a matrix with full column rank, which implies that the selected points are linearly independent.
For the nonnegativity of x_n^{down}, we consider the $N-2$ points a_i^{down} for $i \in \mathcal{N} \setminus \{1, n\}$ and the $N-1$ points a_i^{up} for $i \in \mathcal{N} \setminus \{1\}$ and again choose the points e^{on} and e^{off}. The affine independence can be proved analogously to the previous case.

Inequalities (4.11)

In the following, we focus on an inequality (4.11) for a fixed time step $i \in \{2\ldots,T-L+1\}$ and a selected scenario $s \in \mathcal{S}$ with corresponding path $(v_1,\ldots,v_T)$. Again we proof this inequality to be a facet-defining by constructing $2N-1$ affinely independent binding solutions.

Therefore, we choose the points a_n^{down} for $n \in \{v_2,\ldots,v_{i+L-1}\}$, the points a_n^{up} for $n \in \{v_i,\ldots,v_T\}$ and e^{off}. So far, we have $T+L$ affinely independent points. Additionally, we choose the points a_n^{up} for $n \in \mathcal{N} \setminus \mathcal{N}_s$. Note that the points a_n^{down} with $n \in \mathcal{N} \setminus \mathcal{N}_s$ do not satisfy the inequality by equality.

Dependent upon the scenario s and the minimum runtime L, we define $T-L-1$ additional points $b_n^{s,L}$ for a node $n \in \{v_2,...,v_{T-L}\}$. Here the power plant is switched on in node n and is operating for exactly L consecutive time steps on the path $(v_1,\ldots,v_T)$. This means that the plant is not running in the first node v_1 and in the last node v_T. The state of the plant in the remaining nodes can be chosen such that no additional x_k^{up} or x_k^{down} variable is set to one. More precisely, we construct the point by considering those nodes $k \in \mathcal{N} \setminus \mathcal{N}_s$, for which their predecessor $p(k)$ is in $\mathcal{N}_s$. If $x_{p(k)} = 1$, then we set x_l to one for all nodes $l \in desc(k)$. Analogously, if $x_{p(k)} = 0$, we set x_l to zero for all nodes $l \in desc(k)$.

Finally, we choose $N-T$ points $c_n^{s,L}$ for $n \in \mathcal{N} \setminus \mathcal{N}_s$. Here, the power plant is switched off in node n. If $t(n)-1 \leq L$ the power plant is operating on the $path(p(n))$. In order to create a feasible point which satisfies inequality (4.11) by equality, the power plant is operating on the $path(v_{t(n)-1})$, too. This means that the scenario path $\mathcal{N}_s$ is affected and $x_{v_{t(n)}}^{down} = 1$. Again, all other nodes are chosen such that no additional x_k^{up} or x_k^{down} variable is set to one, as explained above. On the other side, if $t(n)-1 > L$, the plant is running for exactly L time steps on $path(n)$. Then, we distinguish between the following two cases. In the first one, all nodes $k \in path(n)$ with $x_k = 1$ are not an element of $\mathcal{N}_s$, which means that the nodes in $\mathcal{N}_s$ are not affected. Thus, all other nodes can be chosen such that no further x_k^{up} or x_k^{down} variable has the value of one. In the other case, the variable setting on $path(n)$ affects the nodes in $\mathcal{N}_s$. Then we also set $x_{v_{t(n)}}^{down} = 1$ and require that the plant is operating for exactly L nodes on $\mathcal{N}_s$. The variables in all other nodes are chosen as described above.

A graphical example of the point $b_n^{s,L}$ as well as the latter case of point $c_n^{s,l}$ is shown in Figure 4.2. The path corresponding to the scenario s is highlighted in gray. Remember that a black node corresponds to an operating power plant and a white node to a plant which is switched off.

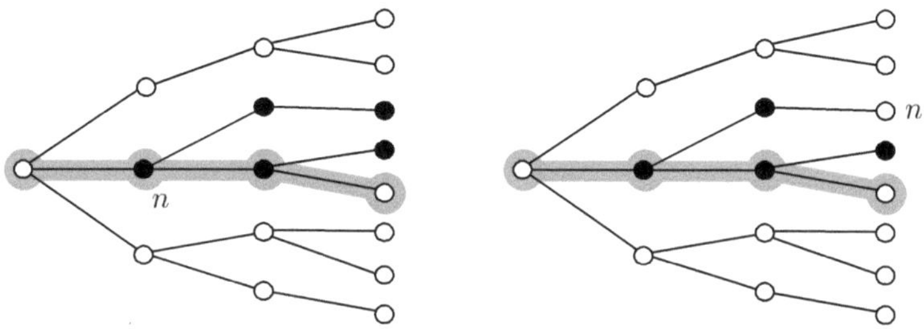

Figure 4.2: Scenario tree representing the points $b_n^{s,L}$ and $c_n^{s,L}$ for $L = 2$

In order to verify the affine independence of the $2N - 1$ points, we subtract the point e^{off} from the other points show that they are linear independent. The resulting matrix is denoted by $M := (M_1, M_2)$, where M_1 corresponds to the first $2T - 2$ columns of the matrix defined by the points $M_1 = (a_{v_i}^{up}, \dots, a_{v_T}^{up}, a_{v_2}^{down}, \dots, a_{v_{i+L-1}}^{down}, b_{v_2}^{s,L}, \dots, b_{v_{T-L}}^{s,L})$. More precisely, these points lead to

$$M_1 = \begin{pmatrix} A_1^{up} & A_2^{up} & A_1^{down} & A_2^{down} & B_1 & B_2 \\ \mathbf{0} & \mathbf{0} & \mathbf{0} & \mathbf{0} & \boldsymbol{I}_{i-2} & \mathbf{0} \\ \boldsymbol{I}_{T-L-i+1} & \mathbf{0} & \mathbf{0} & \mathbf{0} & \mathbf{0} & \boldsymbol{I}_{T-L-i+1} \\ \mathbf{0} & \boldsymbol{I}_L & \mathbf{0} & \mathbf{0} & \mathbf{0} & \mathbf{0} \\ \mathbf{0} & \mathbf{0} & \boldsymbol{I}_L & \mathbf{0} & \mathbf{0} & \mathbf{0} \\ \mathbf{0} & \mathbf{0} & \mathbf{0} & \boldsymbol{I}_{i-2} & \boldsymbol{I}_{i-2} & \mathbf{0} \\ \mathbf{0} & \mathbf{0} & \mathbf{0} & \mathbf{0} & \mathbf{0} & \boldsymbol{I}_{T-L-i+1} \\ A_3^{up} & A_4^{up} & A_3^{down} & A_4^{down} & B_3 & B_4 \\ \mathbf{0} & \mathbf{0} & \mathbf{0} & \mathbf{0} & \mathbf{0} & \mathbf{0} \\ \mathbf{0} & \mathbf{0} & \mathbf{0} & \mathbf{0} & \mathbf{0} & \mathbf{0} \end{pmatrix} \begin{matrix} \left.\begin{matrix} \\ \\ \\ \\ \\ \\ \\ \end{matrix}\right\} \mathcal{N}_s \\ \left.\begin{matrix} \\ \\ \\ \end{matrix}\right\} \mathcal{N} \setminus \mathcal{N}_s \end{matrix}$$

As marked, the first seven lines refer to variables associated with the scenario path s, i.e., the first line corresponds to variables x_i, the second to fourth line to x_i^{up}, and the fifth to seventh line to x_i^{down} with $i \in \mathcal{N}_s$. The last three lines refer to the remaining nodes, respectively, which means to variables x_i, x_i^{up}, and x_i^{down} with $i \in \mathcal{N} \setminus \mathcal{N}_s$. The matrices A_i^{up}, A_i^{down}, and B_i with $i \in \{1, \dots, 4\}$ are zero-one matrices of according dimension. Focusing on the lines two to seven, the submatrix can be transformed to an upper diagonal submatrix containing all columns with ones on the main diagonal. Hence, these $2T-2$ points are linearly independent.

The next $2N-2T$ points of M basically refer to the set $\mathcal{N} \setminus \mathcal{N}_s$. Therefore, we index the elements of $\mathcal{N} \setminus \mathcal{N}_s$ by $\{k_1, \ldots, k_{N-T}\}$. Thus, considering the points $M_2 = (c_{k_1}, \ldots, c_{k_{N-T}}, a^{up}_{k_1}, \ldots, a^{up}_{k_{N-T}})$ leads to

$$\begin{pmatrix} C_1 & A_1 \\ C_2 & A_2 \\ \boldsymbol{I}_{N-T} & \mathbf{0} \\ \mathbf{0} & \boldsymbol{I}_{N-T} \end{pmatrix} \begin{matrix} \}\mathcal{N}_s \\ \\ \left.\begin{matrix} \\ \\ \\ \end{matrix}\right\} \mathcal{N} \setminus \mathcal{N}_s \end{matrix}$$

Here, the first line corresponds to all variables associated with scenario s. The lines two through four represent the variables x_i, x^{up}_i and x^{down}_i with $i \in \mathcal{N} \setminus \mathcal{N}_s$, respectively. Clearly, these points are linearly independent. Observe that in matrix M_1 all entries corresponding to x^{up}_i or x^{down}_i variables with $i \in \mathcal{N} \setminus \mathcal{N}_s$ equal zero. Hence, it follows that all columns of M are linearly independent and altogether we obtain $2N-1$ affinely independent points.

Inequalities (4.12)

Again, let $i \in \{2, \ldots, T-l-1\}$ be fixed and we consider the scenario associated with the path $(v_1, \ldots, v_T)$. For inequality (4.12) we choose the points a^{down}_n for $n \in \{v_i, \ldots, v_T\}$, the points a^{up}_n for $n \in \{v_2, \ldots, v_{i+l-1}\}$ and e^{on}. Additionally, we consider the points a^{down}_n for $n \in \mathcal{N} \setminus \mathcal{N}_s$.

Analogously to the points $b^{s,L}_n$, we specify the $T-l-1$ points $d^{s,l}_n$ for a node $n \in \{v_2, ..., v_{T-l}\}$, for which the power plant is switched off in node n. For the next l consecutive nodes on $\mathcal{N}_s$ the plant is not operating and then switched on again. The x_k variable of all other nodes is set to one or to zero such that no further x^{up}_k or x^{down}_k variable equals one, as explained above.

Further on, we choose the $N-T$ points $f^{s,l}_n$, where the power plant is switched on in node $n \in \mathcal{N} \setminus \mathcal{N}_s$, which means $x^{up}_n = 1$. If $t(n) - 1 \leq l$ the power plant is not operating on the $path(p(n))$. Additionally, it is turned off on the $path(v_{t(n)-1})$ with $x^{up}_{v_{t(n)}} = 1$. This means that the root node and thus, the scenario path $\mathcal{N}_s$ is affected. Again all other nodes are chosen such that no additional x^{up}_k or x^{down}_k variable is set to one, as explained above. On the other side, if $t(n) - 1 > l$, we also distinguish between the following two cases. The first one is that the variable setting on $path(n)$ does not affect the nodes in $\mathcal{N}_s$, namely all variables x_k on the $path(n)$ with $x_k = 0$ are not elements of $\mathcal{N}_s$. Thus, all variables x_k with $k \in \mathcal{N}_s$ can be set to one. Secondly, the variables of the scenario s are affected, which means that at least one x_k with $k \in \mathcal{N}_s$ is forced to zero. These cases can be handled analogously to the two cases described above.

Altogether, we obtain $2N - 1$ affinely independent points, where the affine independence can be shown in analogy to the previous case. □

Yielding a complete linear description of the polytope $P_{\Gamma,L,l}$, in the following lemma we show which of the original constraints are not necessary to linearly describe $P_{\Gamma,L,l}$.

Lemma 4.3 *Let Γ be a scenario tree with the node set $\mathcal{N} = \{1, \ldots, N\}$. The inequalities (4.2), (4.3) and (4.5) are redundant in the system (4.4) and (4.10) to (4.12).*

Proof. Starting with (4.2), we consider a fixed pair $(n, k) \in \mathcal{L}_L^{up}$ and a corresponding scenario $s \in S$, i.e., $n, k \in \mathcal{N}_s$. We know that the system of constraints restricted to Γ_s corresponds to the deterministic case. As described at the beginning of this section, the constraints (4.4) and (4.10) to (4.12) associated with Γ_s provide a complete linear description of the deterministic switching polytope and thus, inequality (4.2) for $n, k \in \mathcal{N}_s$ is redundant. As these constraints describe a subsystem of all inequalities (4.4) and (4.10) to (4.12) associated with the complete scenario tree Γ, inequality (4.2) for $n, k \in \mathcal{N}_s$ is also redundant in the complete system. The inequalities (4.3) and (4.5) can be handled analogously. □

Subsequently, we present the main result of this section, which is the complete description of $P_{\Gamma,L,l}$ by linear inequalities.

Theorem 4.4 *Let Γ be a scenario tree with the node set $\mathcal{N} = \{1, \ldots, N\}$. Equations (4.4) and inequalities (4.10) to (4.12) provide a complete linear description of $P_{\Gamma,L,l}$.*

Proof. Let $Q_{\Gamma,L,l}$ denote the polytope defined by the set of equations (4.4) and inequalities (4.10) to (4.12), i.e.,

$$Q_{\Gamma_s,L,l} = \{x \in \mathbb{R}^{3N-2} \mid x \text{ satisfies conditions (4.4) and (4.10) to (4.12)}\}.$$

In order to prove the theorem, we show that the polytopes $Q_{\Gamma,L,l}$ and $P_{\Gamma,L,l}$ are identical. We know that $P_{\Gamma,L,l} \subseteq Q_{\Gamma,L,l}$ since all equations (4.4) and inequalities (4.10) to (4.12) are valid for $P_{\Gamma,L,l}$.

Now, let $LP_{\Gamma,L,l}$ denote the linear relaxation of the polytope $P_{\Gamma,L,l}$ which is:

$$LP_{\Gamma,L,l} = \{x \in [0,1]^{3N-2} \mid x \text{ satisfies conditions (4.2) to (4.5)}\}.$$

As only valid inequalities are added, see Lemma 4.2, and redundant inequalities are neglected, see Lemma 4.3, we know that $Q_{\Gamma,L,l} \subseteq LP_{\Gamma,L,l}$ holds. It remains to show that $Q_{\Gamma,L,l}$ is integral.

In order to prove the integrality, we show that a fractional point $z \in Q_{\Gamma,L,l}$ cannot be a vertex of $Q_{\Gamma,L,l}$. Therefore, we assume that $\bar{z} \in Q_{\Gamma,L,l}$ is a vertex with at least one fractional component. The idea is to prove that there are not enough linearly independent inequalities in the description of $Q_{\Gamma,L,l}$ that $\bar{z}$ satisfies by equality. Thus, $\bar{z}$ cannot be a vertex, which contradicts our assumption. This is shown by induction on j, which denotes the number of scenarios in the tree.

We begin with the basic case, $j = 1$, where the scenario tree Γ^j contains only one scenario. In this case the polytope $Q_{\Gamma^1,L,l}$ corresponds to the deterministic switching polytope $Q_{T,L,l}$, where T is the number of time steps of Γ^1. By [Mor07], we know that this polytope is integral.

Now, we proceed with $j+1$ scenarios and the corresponding polytope $Q_{\Gamma^{j+1},L,l}$. As inductive hypothesis, we assume that the polytope $Q_{\Gamma^j,L,l}$ with j scenarios is integral. In the following, we describe how to reformulate the description of the polytope $Q_{\Gamma^{j+1},L,l}$, such that it contains the description of $Q_{\Gamma^j,L,l}$.

By $\bar{s}$ we denote the scenario that should be neglected from Γ^{j+1} and $(v_1, \ldots, v_T)$ the corresponding path. Let τ denote the largest number of time steps for which the nodes of $\bar{s}$ coincide with nodes of any other scenario $w \in \mathcal{S} \setminus \{\bar{s}\}$. Denoting the scenario path of w by $(w_1, \ldots, w_T)$, this means that $v_i = w_i$ for all $i \in \{1, \ldots, \tau\}$. To achieve a separate description of the scenario $\bar{s}$, the nodes v_i with $i \in \{1, \ldots, \tau\}$ are duplicated, resulting in a separate one-scenario tree $\Gamma_{\bar{s}}$ and a second scenario tree Γ^j, where $\bar{s}$ is truncated from Γ^j. The corresponding variables and constraints are duplicated, too, resulting in separate description of the polytope $Q_{\Gamma_{\bar{s}},L,l}$ and $Q_{\Gamma^j,L,l}$.

In Figure 4.3, this procedure is graphically described. The selected scenario $\bar{s}$ is highlighted in gray, and the black nodes correspond to the nodes which

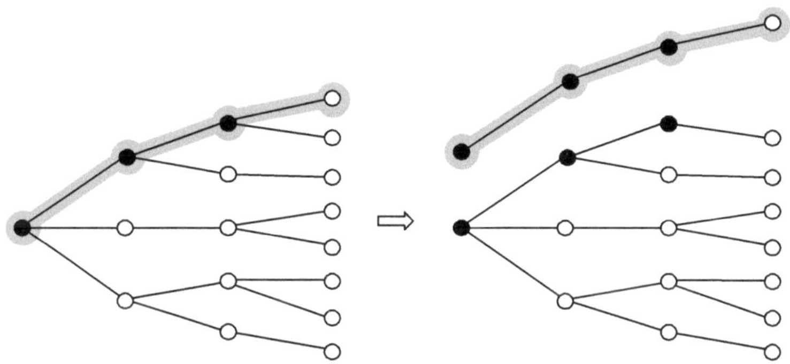

Figure 4.3: Splitting of scenario tree Γ^{j+1} into Γ^j and $\Gamma_{\bar{s}}$ with $\tau = 3$

are duplicated. We remark that in this example the number τ is equal to three.

For a better understanding, we refer to the variables corresponding to the scenario tree Γ^j by

$$x := (x_{u_1}, \ldots, x_{u_N}, x^{up}_{u_2}, \ldots, x^{up}_{u_N}, x^{down}_{u_2}, \ldots, x^{down}_{u_N})^\top,$$

where N is the number of nodes of Γ^j. The variables of the one-scenario tree $\Gamma_{\bar{s}}$ are denoted by

$$y := (y_{v_1}, \ldots, y_{v_T}, y^{up}_{v_2}, \ldots, y^{up}_{v_T}, y^{down}_{v_2}, \ldots, y^{down}_{v_T})^\top.$$

In order to obtain a reformulation of the original polytope, the following $3\tau - 2$ equations are added to the description, ensuring that the duplicated variables take identical values:

$$x_{w_i} - y_{v_i} = 0, \quad \text{for } i \in \{1, \ldots, \tau\}, \tag{4.13}$$

$$x^{up}_{w_i} - y^{up}_{v_i} = 0, \quad \text{for } i \in \{2, \ldots, \tau\}, \tag{4.14}$$

$$x^{down}_{w_i} - y^{down}_{v_i} = 0, \quad \text{for } i \in \{2, \ldots, \tau\}. \tag{4.15}$$

Remember that w_1 to w_τ describe the duplicated nodes of Γ^j. Additionally, we remark that the reformulated polytope $\bar{Q}_{\Gamma^{j+1},L,l} \subseteq \mathbb{R}^{3N-2+3T-2}$.

Now, let y be sorted such that the variables associated with the nodes v_1 to v_τ are the first elements of y and the variables of x are sorted accordingly, with respect to the nodes w_1 to w_τ. Transforming each equation (4.4) into two inequalities, the switching polytope with $j+1$ scenarios can be described as

$$\begin{aligned}\bar{Q}_{\Gamma^{j+1},L,l} = \{\tbinom{x}{y} \in \mathbb{R}^{3N-2+3T-2} |\ & Ax \leq c \\ & By \leq d \\ & (\boldsymbol{I}_{3\tau-2}, \boldsymbol{0})x - (\boldsymbol{I}_{3\tau-2}, \boldsymbol{0})y = 0\,\}\end{aligned} \tag{4.16}$$

where $A \in \mathbb{R}^{4N-4+j(2T-L-l)\times 3N-2}$, $B \in \mathbb{R}^{6T-L-l-4\times 3T-2}$ and $\boldsymbol{I}_{3\tau-2}$ is the identity matrix of dimension $3\tau-2$. In detail, the system $Ax \leq c$ consists of $2N-2$ inequalities resulting from equations (4.4), $2N-2$ nonnegativity constraints (4.10) and $j(2T-L-l)$ inequalities (4.11) and (4.12).Thus, we can write $Q_{\Gamma^j,L,l} = \{x \in \mathbb{R}^{3N-2} | Ax \leq c\}$ and $Q_{\Gamma_{\bar{s}},L,l} = \{y \in \mathbb{R}^{3T-2} | By \leq d\}$. This completes the reformulation of $Q_{\Gamma^{j+1},L,l}$ to $\bar{Q}_{\Gamma^{j+1},L,l}$, which includes the description of $Q_{\Gamma^j,L,l}$.

Now, we assume that $\bar{z} \in \bar{Q}_{\Gamma^{j+1},L,l}$ is a vertex with fractional components. In order to describe a vertex of $\bar{Q}_{\Gamma^{j+1},L,l}$, we need $3N+3T-4$ linearly independent inequalities, as $\bar{Q}_{\Gamma^{j+1},L,l} \subseteq \mathbb{R}^{3N+3T-4}$. The matrix corresponding to the formulation (4.16) reduced to the equality set of $\bar{z} = \tbinom{\bar{x}}{\bar{y}}$ can be written as

$$M = \begin{pmatrix} C \\ D \\ I \end{pmatrix} \qquad \text{with} \qquad \begin{aligned} C &= \begin{pmatrix} A_{eq(\bar{x})\cdot}, & \boldsymbol{0} \end{pmatrix}, \\ D &= \begin{pmatrix} \boldsymbol{0}, & B_{eq(\bar{y})\cdot} \end{pmatrix}, \\ I &= \begin{pmatrix} \boldsymbol{I}_{3\tau-2}, & \boldsymbol{0}, & -\boldsymbol{I}_{3\tau-2}, & \boldsymbol{0} \end{pmatrix}. \end{aligned}$$

where $eq(\bar{x})$ denotes the equality set of $\bar{x}$, i.e., it contains the indices of all those rows of A whose corresponding constraints are tight for $\bar{x}$. Set $eq(\bar{y})$ is explained analogously. Assuming that $\bar{z}$ is a vertex, the rank of M has to equal $3N + 3T - 4$, as explained above. By construction, there is a $g \in \{0, \ldots, 3N - 2\}$ with $\text{rank}(A_{eq(\bar{x})\cdot}) = 3N - 2 - g$. Analogously, there exists an $h \in \{0, \ldots, 3T-2\}$ so that the equation $\text{rank}(B_{eq(\bar{y})\cdot}) = 3T-2-h$ holds. Finally, we know that $\text{rank}(I_{3\tau-2}, \boldsymbol{0}, -I_{3\tau-2}, \boldsymbol{0})$ equals $3\tau - 2$. Based on our inductive hypothesis, g and h can only be zero if $\bar{x}$ and $\bar{y}$ are integral. Note that $g = h = 0$ implies that $\bar{x}$ and $\bar{y}$ are vertices of the polytopes $Q_{\Gamma^j,L,l}$ and $Q_{\Gamma_{\bar{s}},L,l}$, respectively. Additionally, we know that the polytopes $Q_{\Gamma^j,L,l}$ and $Q_{\Gamma_{\bar{s}},L,l}$ are integral, applying our assumption. Hence, assuming $\bar{z} = \tbinom{\bar{x}}{\bar{y}}$ to be fractional, we can deduce that $g + h \geq 1$.

In order to contradict our assumption that $\text{rank}(M) = 3N + 3T - 4$, i.e., $\bar{z}$ is a vertex, we distinguish the following two cases:

$\boxed{g + h > 3\tau - 2}$

Since $\text{rank}(I) = 3\tau - 2$, it follows directly that

$$\text{rank}(M) = \text{rank}(C) + \text{rank}(D) + \text{rank}(I) < 3N + 3T - 4,$$

which is a contradiction to our assumption.

$\boxed{1 \leq g + h \leq 3\tau - 2}$

As $\text{rank}(C) = 3N - 2 - g$, we can reduce the matrix C to $3N - 2 - g$ rows, such that the matrix has full row rank. Analogously, the matrix D is reduced to $3T - 2 - h$ rows. Thus, by elementary row operations, the matrices C and D can be transformed to

$$C' = \left(\begin{array}{cccccc} c'_1 & a'_{11} & & & & \\ \vdots & \vdots & \ddots & & \mathbf{0} & \mathbf{0} \\ c'_G & a'_{G,1} & \dots & a'_{G,G} & & \end{array} \right), \quad \underbrace{}_{g} \; \underbrace{\phantom{a'_{G,1} \dots a'_{G,G}}}_{3N-2-g} \; \underbrace{\phantom{\mathbf{0}\;\mathbf{0}}}_{3T-2}$$

where $G = 3N - 2 - g$ and $(c'_i)^\top \in \mathbb{R}^g$ and

$$D' = \left(\begin{array}{cccccc} & & d'_1 & b'_{11} & & \\ \mathbf{0} & \mathbf{0} & \vdots & \vdots & \ddots & \\ & & d'_H & b'_{H,1} & \dots & b'_{H,H} \end{array} \right), \quad \underbrace{\phantom{\mathbf{0}\;\mathbf{0}}}_{3N-2} \; \underbrace{}_{h} \; \underbrace{\phantom{b'_{H,1} \dots b'_{H,H}}}_{3T-2-h}$$

where $H = 3N - 2 - h$ and $(d'_i)^\top \in \mathbb{R}^h$.

In the following, we distinguish the following two cases, where either $\bar{x}$ or $\bar{y}$ is not integral.

<u>**Case** 1 ($\bar{y}$ is fractional):</u>

First, we assume that an element of $\bar{y}$ is fractional. Using the matrix I, the matrix C' can be transformed to

$$C'' = \left(\begin{array}{cccccccc} & & & -c'_1 & -a'_{1,1} & & & \\ \mathbf{0} & \mathbf{0} & \mathbf{0} & \vdots & \vdots & & \ddots & \mathbf{0} \\ & & & -c'_{\phi-g} & -a'_{\phi-g,1} & \cdots & -a'_{\phi-g,\phi-g} & \\ c'_{\phi-g-1} & a'_{\phi-g+1,1} & \cdots & & & & & \\ \vdots & \vdots & \ddots & \mathbf{0} & & & \mathbf{0} & \mathbf{0} \\ c'_G & a'_{G,1} & \cdots \; \cdots \; a'_{G,G} & & & & & \end{array} \right),$$

with underbraces below the columns labelled ϕ, $3N-2-\phi$, ϕ, $3T-2-\phi$

where $\phi = 3\tau - 2$. Here, the first $\phi - g$ lines are transformed using the first $\phi - g$ lines of I. Remember that I connects the first $3\tau - 2$ components of $\bar{x}$ with the first $3\tau - 2$ components of $\bar{y}$. Further on, we mark that the remaining lines of C'' stay unchanged.
Let C''_1 denote the matrix associated with the first $\phi - g$ rows of C'', and the remaining rows are denoted by C''_2. As $\bar{y}$ is not integral, we know that $\operatorname{rank}\begin{pmatrix} C''_1 \\ D' \end{pmatrix} < 3T - 2$. Note that C''_1 also describes the point $\bar{y}$. Additionally, we know that $\operatorname{rank}(C''_2) \leq 3N - 2 - \phi = 3N - 3\tau$. Thus, we obtain

$$\operatorname{rank}(M) \leq \operatorname{rank}(C''_2) + \operatorname{rank}\begin{pmatrix} C''_1 \\ D' \end{pmatrix} + \operatorname{rank}(I) < 3N + 3T - 4,$$

which is a contradiction to our assumption.

Case 2 ($\bar{x}$ is fractional):

If we assume that an element of $\bar{x}$ is fractional, the matrix D' can be transformed, analogously, and the matrix C' stays the same. Hence, this case can be proved in analogy to the first case.

Altogether we proved that the polyhedron $Q_{\Gamma_{S+1},L,l}$ with $S+1$ scenarios is integral, which completes the induction and the proof of the theorem. □

4.4 Separation

In this section we discuss the use of the presented inequalities (4.10) to (4.12) in form of cutting planes within branch-and-cut procedures in order to enhance the solution of the stochastic problem described in Section 3.2.2. First, we remark that for each power plant $i \in \mathcal{I}$ we need $N-1$ equations, $2N-2$ nonnegativity constraints and $S(2T-L-l)$ inequalities in order to describe the subpolytope $P_{\Gamma,L,l}$ completely by linear constraints. Remember that S denotes the number of scenarios of the corresponding scenario tree and T the number of time steps.

At first view, the small number of constraints suggests their incorporation into the original model formulation. However, these additional inequalities could slow down the solution process as in each node of a branch-and-cut procedure, which is based on LP-relaxation, the corresponding linear problem has to be solved. Hence, we also consider the possibility of using these inequalities as cutting planes within the branch-and-cut procedure, which means adding violated inequalities successively during the branch-and-cut process. Especially for the solution of the *S-OPGen* problem, this approach is stressed by the observation that coal power plants are switched on or off infrequently, as energy storages are used to buffer fluctuating supply and demand. As a consequence, minimum run time and down time conditions are rarely violated. Thus, we look at the separation problem associated with these inequalities.

We start with the presentation of a separation algorithm associated with the constraints (4.11) which proceeds as follows. Given a fractional point $x = (x_1, \dots, x_N, x_1^{up}, \dots, x_N^{up}, x_1^{down}, \dots, x_N^{down})^\top$, the most violated inequality is returned. The procedure is an adapted version of the separation algorithm for the deterministic case, presented in [Mar05]. In the algorithm, we use the notation $(v_1^s, \dots, v_T^s)$ to refer to the nodes associated with scenario s.

Basically, Algorithm 4.5 searches iteratively through all constraints for the maximum violation. In the first line we initialize the variable Δ^{max} by zero, which represents the maximum violation detected during the execution and s^{max} and i^{max} denote the corresponding scenario and time step, respectively. The outer for loop iterates over all scenarios. In line three, the violation Δ corresponding to constraint (4.11) with $i = T - L + 1$ of the current scenario s is computed. If no cut with higher violation was detected before, we update Δ^{max}, s^{max}, and i^{max} in line five. In the inner loop, we iterate over all time step i starting with $i = T - L$ and reducing i by one

Algorithm 4.5 *Separation algorithm for inequalities (4.11)*

Input: Inequalities (4.11) and a fractional point x
Output: Pair (s^{max}, i^{max}) defining a maximally violated inequality by x

set $\Delta^{max} = 0$, $s^{max} = 0$, and $i^{max} = 0$
for $s = 1$ to S **do**
  compute $\Delta = -x_{v_T^s} + \sum_{k=T-L+1}^{T} x_{v_k^s}^{up}$
  if $\Delta > \Delta^{max}$ **then**
    set $\Delta^{max} = \Delta$, $s^{max} = s$ and $i^{max} = T - L + 1$
  end
  for $i = T - L$ to 2 **do**
    compute $\Delta = \Delta + x_{v_i^s}^{up} - x_{v_{i+L}^s}^{down}$
    if $\Delta > \tilde{\Delta}$ **then**
      set $\Delta^{max} = \Delta$, $s^{max} = s$ and $i^{max} = i$
    end
  end
  return indices s^{max} and i^{max}
end

until $i = 2$. For each constraint corresponding to the pair (s, i), the violation is represented by Δ again. Indeed, as all constraints of type (4.11) are considered during the procedure, Algorithm 4.5 returns the most violated constraint indicated by s^{max} and i^{max}.

A separation algorithm corresponding to inequality (4.12) can be formulated, analogously. Both of them have a running time of $\mathcal{O}(ST)$. Concerning the nonnegativity constraints, a separation algorithm can be formulated with running time $\mathcal{O}(N)$ by iterating over all nodes $n \in \mathcal{N}$ where $N = |\mathcal{N}|$ is the number of nodes of the scenario tree. Thus, determining the most violated inequality of a point x out of all inequalities (4.10) to (4.12) has a running time of $\mathcal{O}(ST)$, as always $N \leq ST$ holds.

Finally, we refer to Section 8.2 where the separation procedures are computationally investigated and compared to the version where all constraints are added explicitly to the original model.

Chapter 5

Primal Heuristics

So far, we have mainly concentrated on improving the formulation of the problem yielding a better lower bound of the linear program relaxation. A further important aspect in a branch-and-cut algorithm is the generation of good feasible solutions early in the solution process with the aim of reducing the overall computational effort. Thus, in this chapter we focus on the development of a primal heuristic, aiming at the generation of solutions with a low objective function value in an adequate running time. ding the deterministic as well as the stochastic problem, a variety of primal approaches can be found in the literature, generally classified into construction and improvement heuristics. In order to obtain a good feasible start solution for the branch-and-cut algorithm, we follow the idea of *relax-and-fix*, which constructs a feasible solution from scratch. Thereon, we adapt this approach to our problems by developing problem specific approximation schemes, which are used additionally to the integrality relaxation.

After giving a short overview of related literature, we present the general idea of the relax-and-fix heuristic in Section 5.1. As this approach can be applied to the deterministic as well as to the stochastic problem, afterwards we describe how it can be tailored to both problems. With regard to the deterministic case, we call this approach *rolling horizon*, which is described in Section 5.2. Finally, we present the adaptation to the stochastic problem in Section 5.3.

5.1 Relax-and-Fix

The *relax-and-fix* algorithm is a construction heuristic, developed for large-scale mixed-integer programs. It is based on the approach of decomposing a large problem into several smaller ones, which are solved iteratively to generate a feasible solution of the original problem, see [Wol98]. There is a variety of contributions, where *relax-and-fix* is applied to different kind of problems, for instance in air traffic flow management, e.g. in [AAEO00], and project scheduling, see e.g. [ES05]. Especially, concerning lot sizing problems, this heuristic was successfully applied in [DEWZ94]. As the *OP-Gen* problem presented in Chapter 3 shows certain similarities to production planning problems, this contribution additionally motivates the investigation of relax-and-fix within our framework.

The basic procedure of this approach starts with a subdivision of all integer variables into subsets. Based on these subsets, simplified subproblems are generated, where integrality is only required for the currently chosen subset. In particularly, at each time step those variables related to the previously considered subset are frozen according to the solution obtained, and the remaining binary variables are relaxed. Thus, solving the resulting problems iteratively, a feasible solution of the original problem is determined by progressively composing the decisions obtained from solutions of the previous subproblems. Indeed, the solution satisfies all integrality restrictions at the end of the procedure and thus is feasible. As the involved variables and constraints of the subproblems are reduced, the relax-and-fix procedure enables to find good feasible solutions of large scale problems. For the description of the basic concept, we follow the notation of [PW06].

In detail, the heuristic can be described as follows. Let the original problem P be formulated by

$$
(P) \qquad \begin{aligned}
\min \quad & c^\top x + d^\top y \\
\text{s.t.} \quad & Ax + By \geq b \\
& x \in \mathbb{R}^n_+ \\
& y \in \{0,1\}^p
\end{aligned}
$$

given $c \in \mathbb{R}^n, d \in \mathbb{R}^p, A \in \mathbb{R}^{m \times n}, B \in \mathbb{R}^{m \times p}$, and $b \in \mathbb{R}^m$. Now, we assume that the indices of the integer variables are partitioned into R disjoint sets $\mathcal{I}_r$ with $r \in \mathcal{R} := \{1, \dots, R\}$, preferably sorted with decreasing importance. In each iteration r, a reduced problem P_r is generated and solved, where integrality is only imposed on the integer variables associated with set $\mathcal{I}_r$. In iteration $r = 1$, this means that the integrality restrictions of all binary variables are relaxed except for those associated with $\mathcal{I}_1$, resulting in the reduced problem P_1. Let $(\tilde{x}^1, \tilde{y}^1)$ denote an optimal solution of P_1, then in the next iteration all variables y_i with $i \in \mathcal{I}_1$ are fixed to the values of the corresponding solution $\tilde{y}^1$. Transferred to iteration r, we denote an optimal

---

Algorithm 5.1 *Relax-and-Fix Algorithm*

Step 1: Initialization
Partition the index set $\mathcal{I}$ into R subsets $\mathcal{I}_r$. Set $r = 1$.

Step 2: Solving the first problem
Generate problem P_1 with respect to the index set $\mathcal{I}_1$ and solve it.
If P_1 is infeasible, stop and return "problem P is infeasible".

Step 3: Stopping
If $r = R$, stop and return "problem P is feasible".
Otherwise, set $r = r + 1$.

Step 4: Solving problem $\boldsymbol{P_r}$
Generate problem P_r with respect to the index set $\mathcal{I}_r$ and solve it.
If P_r is infeasible, stop and return "status of problem P is unknown".
Otherwise, go to Step 3.

---

solution of P_r by $(\tilde{x}^r, \tilde{y}^r)$. Thereupon, the problem P_r for $r \in \mathcal{R}\backslash\{1\}$ is described by

$$(P_r) \qquad \begin{array}{rll} \min & c^\top x + d^\top y & \\ \text{s.t.} & Ax + By \geq b & \\ & x \in \mathbb{R}^n_+ & \\ & y_i = \tilde{y}_i^{r-1} & \text{for all } i \in \mathcal{I}_1 \cup \ldots \cup \mathcal{I}_{r-1} \\ & y_i \in \{0,1\} & \text{for all } i \in \mathcal{I}_r \\ & y_i \in [0,1] & \text{for all } i \in \mathcal{I}\backslash\{\mathcal{I}_1 \cup \ldots \cup \mathcal{I}_r\} \end{array}$$

where $\mathcal{I} = \cup_{r\in\mathcal{R}}\mathcal{I}_r$. The basic algorithmic framework of relax-and-fix is described in Algorithm 5.1, following the description of [ES05].

Note that in iteration $r = 1$, the optimal objective function value of P_1 provides a lower bound on the optimal function value of P as no integer variable is fixed yet and thus P_1 is a relaxation of P. Clearly, this is not valid for problems P_r with $r \geq 2$. If all subproblems are feasible, we obtain a feasible solution for the original problem in a finite number of time steps.

Nevertheless, there may occur infeasible problems P_r during the process, even if the original problem is feasible. In order to avoid the failure of the complete procedure, several approaches are developed. One possibility is to suspend certain fixations, as proposed by [BGGG06], who include a backward grouping step into the procedure. In particular they redefine the partitioning structure by setting $\mathcal{I}_{r-1} = \mathcal{I}_r \cup \mathcal{I}_{r-1}$ if the current problem P_r is infeasible. Then, P_{r-1} is resolved, imposing integrality on variables associated with the enlarged set $\mathcal{I}_{r-1}$. Note that in the worst case, this results in solving the complete original problem. A further possibility is to add additional constraints in advance, which should avoid running into infeasible reduced problems. Clearly, these constraints can only be generated for specific problems, yielding preferably valid constraints. But constituting a heuristic approach, these constraints do not have to be valid necessarily, see for instance [PW06].

Based on the relax-and-fix framework presented above, a lot of variants of this approach exist. Particularly, the decomposition of the variables is often connected to structural characteristics of the specific problem, as for instance the machine type in a lot sizing problem or a time connected decomposition for time-dependent problems. The latter approach we also call rolling horizon, which is explained in the following section.

5.2 A Rolling Horizon Approach to the *D-OPGen* Problem

In the following, we present a primal heuristic based on the idea of a rolling horizon, which is a modification of the relax-and-fix heuristic explained above. The main idea is the decomposition of the planning horizon $\mathcal{T}$ into several periods. Based on these periods, where the problem is formulated exactly, simplified subproblems are generated. As mentioned above, the progressive solution of the resulting problems allows the consideration of the decisions already made as well as a foresight into the future. This concept was already studied in the seventies, where [Bak77] investigated the effectiveness of rolling horizon decision making in production planning. A survey on related literature can be found in [CHS02].

Compared to the relax-and-fix approach presented above, the major differences of the rolling horizon method applied to the *D-OPGen* problem concern the freezing of variables, the overlapping time windows, and the approximation scheme. Focusing on the specific modifications and on the adapted notation, the heuristic is described in the following.

In this framework, a subproblem P_r in iteration r is based on the subdivision of all time steps $\mathcal{T} = \{1, \ldots, T\}$. This means that in contrast to the subdivision of all integer variables described above, here, subsets $\mathcal{T}_r \subseteq \mathcal{T}$ are

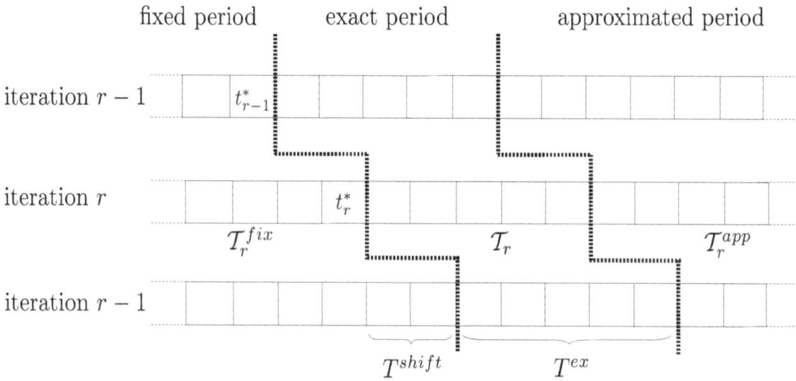

Figure 5.1: Subdivisions of the planning horizon

created. Yielding a uniform subdivision of the planning horizon, we pose the following requirement on the construction of $\mathcal{T}_r$

$$\mathcal{T}_r = \{t \in \mathcal{T} \mid t - T^{shift} \in \mathcal{T}_{r-1}\} \tag{5.1}$$

for all $r = \{2, \dots, R\}$. Here, subset $\mathcal{T}_1$ contains the first steps of the planning horizon, whose number is denoted by T^{ex}. This means that in iteration $r+1$, the current subset $\mathcal{T}_r$ is set back T^{shift} time steps, yielding the new subset $\mathcal{T}_{r+1}$. Consequently, the subsets $\mathcal{T}_r$ are not necessarily disjoint, depending on T^{shift}. Note that only T^{ex} and T^{shift} need to be initialized for a well-defined subdivision of $\mathcal{T}$. Thereon, we define the two sets

$$\mathcal{T}_r^{fix} = \{t \in \mathcal{T} \mid t < \underline{t}_r\}, \tag{5.2}$$

$$\mathcal{T}_r^{app} = \{t \in \mathcal{T} \mid t > \bar{t}_r\}, \tag{5.3}$$

for all $r = \{1, \dots, R\}$, where $\underline{t}_r$ and $\bar{t}_r$ denote the first and the last time step of $\mathcal{T}_r$, respectively.

Now, we are able to describe P_r in detail. A subproblem in iteration r is basically composed of three periods. The first one is referred to as the *fixed period*, described by the set $\mathcal{T}_r^{fix}$. The set of variables associated with this period is fixed, retrieving the values of the variables from previous iterations. Note that in contrast to relax-and-fix, we freeze all kinds of variables corresponding to these time steps, i.e., binary and continuous variables. The second part is named *exact period* and consists of the time steps for which the problem is formulated exactly. The time steps corresponding to

Algorithm 5.2 *Rolling Horizon Algorithm*

Step 1: Initialization
Initialize parameters T^{ex} and T^{shift} and construct subsets $\mathcal{T}_1$ to $\mathcal{T}_R$ according to (5.1) and set $r = 1$.

Step 2: Solving the first problem
Generate problem P_1 by approximating the problem for time steps $t \in \mathcal{T}_1^{app}$ defined in (5.2) according to a chosen relaxation or approximation strategy and solve it.
If P_1 is infeasible, stop and return "status of problem P unknown".

Step 3: Stopping
If $r = R$, stop and return "problem P is feasible".
Otherwise, set $r = r + 1$.

Step 4: Solving problem P_r in iteration r
Fixed period. The variables corresponding to the time steps $t \in \mathcal{T}_r^{fix}$ are fixed, according to the solution of P_{r-1}.
Exact period. Considering the time steps $t \in \mathcal{T}_r$, the problem is formulated exactly.
Approximated period. Considering the time steps $t \in \mathcal{T}_r^{app}$, the problem is described according to a chosen relaxation or approximation strategy.
Solve the problem P_r.
If P_r is infeasible, stop and return "status of problem P is unknown".
Otherwise, go to Step 3.

---

this period are contained in $\mathcal{T}_r$. Finally, the last time steps are combined in the *approximated period* associated with the set $\mathcal{T}_r^{app}$. Based on a predefined approximation strategy, a relaxed or approximated formulation of the problem restricted to these time steps is assumed. The approximation strategy enables us to include future events in our present decisions without solving the original problem. Alternatively, parts of this period can also be neglected. Figure 5.1 graphically illustrates the procedure. In summary, the rolling horizon method is described in Algorithm 5.2, which is an adaptation and specification the Relax-and-Fix Algorithm 5.1 of the previous section.

We remark that using approximation strategies instead of relaxing integrality, problem P_1 is no longer a relaxation of P. Consequently, P does not have to be infeasible, if the solution status of P_1 is infeasible.

The choice of the parameters T^{shift} and T^{ex} considerably influences the performance of the heuristic. On the one hand, comprising a great number of time steps in T^{ex}, a lot of detailed information can be used as less time steps are described approximately. On the other hand, a great value of T^{ex} means more complex subproblems P_r. Concerning the parameter T^{shift}, a small value decreases the restrictive impact of the fixed section. However, T^{shift} is directly linked to the total number of iterations of the algorithm. In conclusion, choosing these parameters, the quality of the solution and the computational costs have to be thoroughly balanced. For the *D-OPGen* problem described in Section 3.1.6, these parameters are chosen by means of a series of test runs, see Section 8.3.1.

In the following two sections, the adaptation of the algorithm to our problem is described. At first, we discuss the treatment of the approximated period and afterwards investigate the feasibility of a subproblem P_r.

5.2.1 Approximation Strategies

Next to the classical integrality relaxation, we present two further approximation strategies which are based on the characteristics of our problem. The aim of both strategies is the reduction of the computational costs, maintaining as much future information as possible. This approach is motivated by the observation made during a number of test runs that events which lie far in the future have less impact on current decisions than those in the near future.

Integrality Relaxation R

An intuitive approach of handling the modeling in times steps corresponding to the approximated period $\mathcal{T}_r^{app}$ is the relaxation of the integrality restriction, as presented in the original relax-and-fix algorithm in Section 5.1. With respect to these time steps, this approach provides the advantage of resulting in a pure LP-formulation in contrast to the following approximation strategies.

Approximation Strategy S_1

The first approximation strategy addresses the piecewise linear efficiency functions $f(x)$, occurring within the description of the power plants and energy storages, see Section 3.1.5. Remember that all piecewise linear functions approximate nonlinear terms which include an efficiency function $\eta(x)$.

More precisely, there are the following two types of nonlinear terms which are approximated:

$$f(x) \approx x \cdot \eta(x) \quad \text{and} \quad f(x) \approx \frac{x}{\eta(x)}. \tag{5.4}$$

The variable x represents the charged or discharged power of the energy storages or the produced power of the power plants, respectively. The basic idea of this strategy is the approximation of each of the nonlinear terms by one linear function instead of the piecewise linear approximation function $f(x)$. For this purpose we present two alternative approaches.

The first possibility, called $\bar{S}_1$, approaches the nonlinear relation by approximating the efficiency function $\eta(x)$ by a constant efficiency $\bar{\eta}$. Applying the least square method in the points $x_i \in [x^{min}, x^{max}]$ which are also used for the piecewise linear approximation, we obtain a linear function $g(x) = \bar{\eta}x$ or $g(x) = \frac{x}{\bar{\eta}}$, respectively. Thus, a piecewise linear approximation of the terms is no longer necessary, as they only depend linearly on the variable x. This is a common approach simplifying the description of such technical relations, see e.g. [CS98].

The second possibility $\tilde{S}_1$ yields a closer approximation of the nonlinear terms (5.4). Again the least square method in the grid points of $f(x)$ is applied, yielding a linear function $h(x) = ax + b$ for this approximation strategy. Naturally, comparing the method using constant efficiencies with this one, the latter causes a smaller approximation error in the grid points. But note that in case the function $h(x)$ is used, the corresponding binary decision variable, indicating the state of the approximated process, has to be involved in order to ensure that the function takes the value zero if x equals zero. In contrast, fixing the efficiency to $\bar{\eta}$, this condition is already satisfied. Figure 5.2 illustrates a comparison of both strategies exemplarily for the charging function of the pumped hydro storage.

Altogether, using this approximation strategy, the number of binary variables of each subproblem is drastically reduced, as all binary variables needed for the piecewise linear approximation are completely neglected, which constitute the major part of all binary variables.

Approximation Strategy S_2

This strategy yields a coarsening of the problem description by lowering the time resolution within this time window. This means that a certain number of time steps is aggregated to one time step. For example, the original resolution of 15 minutes is coarsened to an hourly time resolution, which

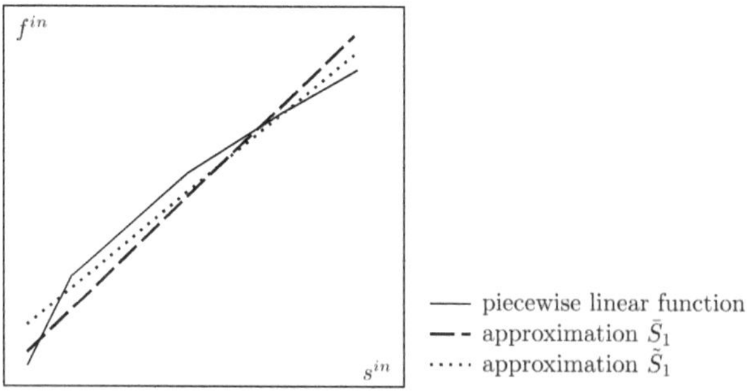

Figure 5.2: Approximations of a charging function

means that only a fourth of the original variables are needed to describe the problem associated with this time window. Note that within the new time resolution the binary decision variables as well as the piecewise linear functions are maintained.

Furthermore, there arises the possibility of coupling these two approximation strategies. For example, the time steps can be aggregated according to strategy S_2 and afterwards the linear approximation of the piecewise linear functions can be applied as described in strategy S_1. This results in a further reduction of the problem size.

Extended Relaxation R^* and Approximations S^*

Finally, we present a variant where only the first part of the time window is considered and the last part is completely neglected. This means that only a certain number of time steps, which we denote by T^{app}, are approximated by R, S_1 or S_2. The set of corresponding time steps is defined by

$$\mathcal{T}^{app^*} = \{t \in \mathcal{T}_r^{app} | t > \bar{t}_r + T^{app}\}.$$

Remember that $\bar{t}_r$ is the last time step of the exact period $\mathcal{T}_r$ in iteration r. All time steps $t \in \mathcal{T}^{app} \setminus \mathcal{T}^{app^*}$ are completely omitted for this iteration step. This extension is denoted by R^*, S_1^* and S_2^*, respectively. Especially for long planning horizons this approach is reasonable assuming that events far in the future hardly ever influence present decisions. As in case of the parameters T^{shift} and T^{ex}, the choice of T^{app} affects the running time of the algorithm as well as the quality of the solution obtained by the heuristic. Thus, it is also chosen by means of a series of test runs, see Section 8.3.1. The advantage of this approach is that for a fixed T^{app} the size of the subproblems P_r does not increase for longer planning horizons T. Note that additionally the number of subproblems only rises linearly with T, and therefore larger problem instances can be considered.

5.2.2 Feasibility

The next section deals with the feasibility of the subproblems P_r under the assumption that the original problem is feasible. As in each iteration r of the rolling horizon algorithm only a limited number of time steps are modeled in detail, i.e., those corresponding to the exact period, the occurrence of an infeasible subproblem P_r for an $r \in \mathcal{R}$ cannot be excluded. Concerning the *D-OPGen* problem, the infeasibility may result from already fixed storage levels which are too low in order to achieve the predefined storage level at the end of the planning horizon. By sharpening the restrictions on the storage level s_{jt} of storage $j \in J$ in time step $t \in \mathcal{T}$, this problem can be diminished. More precisely, the level s_{jt} has to be high enough to be capable to reach the final storage level $\bar{s}_{jT}$, charging at full capacity. Taking the start-up energy α_j^{in} for charging into account, we obtain the following inequality

$$s_{jt} \geq \max\{s_j^{min}, \bar{s}_{jT} - \frac{T-t}{\tau} \sum_{k \in \mathcal{K}_j} f_k^{in,max} + \alpha_j^{in}(1 - y_{jt}^{in})\}, \tag{5.5}$$

for all $j \in \mathcal{J}$ and $t \in \mathcal{T} \setminus \{T\}$. Here, $f_k^{in,max}$ denotes the maximum function value of the piecewise linear function $f_j^{in}(s_{kt}^{in})$. Hence, the product of $T-t$ with the maximum amount of charging energy per times step $\frac{1}{\tau}\sum_{k\in\mathcal{K}_j} f_k^{in,max}$ describes the maximum overall energy possibly charged until the end of the planning horizon. Under the assumption that the original problem is feasible, the sharpened storage level bounds may avoid to run into infeasible subproblems P_r. If all P_r are feasible, the heuristic provides a feasible solution in finitely many iterations, which naturally may be suboptimal.

Imposing certain requirements on the parameter set of the *D-OPGen* problem, we are able to ensure feasibility of the subproblems. Clearly, the feasibility of P_r highly depends on the approximation used for the time steps in $\mathcal{T}_r^{app}$. Therefore, we start by considering the integrality relaxation R for the approximated period.

Under this assumption, basically the charging process may cause infeasibilities, as this operation need to be performed if the storage level falls below the final storage level $\bar{s}_{jT}$ during the planning horizon. Note that in contrast, the discharging operation is not necessary in order to generate a feasible solution, as only a lower bound on the final storage level is given. Hence, we require

$$s_k^{in,min} = 0, \tag{A}$$

for all $k \in \mathcal{K}_j$ and $j \in \mathcal{J}$, which enables the charging of arbitrarily small amounts of power for all energy storages $j \in \mathcal{J}$. Thus, imposing (A) on the parameter set of the *D-OPGen* problem, we obtain the following result. We remark that in the following theorem, we abbreviate the *D-OPGen* problem by P.

Theorem 5.3 *Let conditions* (A) *be fulfilled by the parameter set of problem P and assume inequalities (5.5) are added to the description of P. Using integrality relaxation R for the approximated period, subproblem P_r in each iteration r of the rolling horizon algorithm is feasible if P is feasible.*

Proof. Except for the minimum runtime and downtime restrictions (3.4) and (3.5) , we know that in the problem formulation at most two consecutive time steps are connected. Neglecting constraints (3.4) and (3.5), consequently the feasibility of the subproblem P_r in iteration r only depends on the values of the frozen variables corresponding to the last time step t_r^* of the fixed period $\mathcal{T}_r^{fix}$, which is marked in Figure 5.1. As integrality relaxation is assumed for the approximated period $\mathcal{T}_r^{app}$, the reformulation of this period does not restrict the feasibility of P_r. Hence, only if the variables of time step t_r^* are chosen inappropriately, P_r becomes infeasible.

In the following, we show the feasibility of P_r by specifying the construction of a feasible solution based on the fixed variables in time step t_r^*. We start with iteration $r = 1$. As problem P_1 provides a linear relaxation of the original problem, it is always feasible.

Now, we consider iteration $r > 1$, where $\mathcal{T}_r^{fix} \neq \emptyset$, i.e., there exists at least one time step $t \in \mathcal{T}$ for which the corresponding variables are fixed. We start with constraint (3.1) concerning the fulfillment of demand. As the imported power x_t is unbounded, for each $t \in \mathcal{T}_r \cup \mathcal{T}_r^{app}$ the variable can be set to a value large enough such that this constraint is satisfied. In detail, using constraint (3.1), we set

$$\bar{x}_t = -\sum_{i \in \mathcal{I}} \bar{p}_{it} - \sum_{j \in \mathcal{J}} \sum_{l \in \mathcal{L}_j} \bar{s}_{lt}^{out} + \sum_{j \in \mathcal{J}} \sum_{k \in \mathcal{K}_j} \bar{s}_{kt}^{in} - \omega_t + \delta_t,$$

where $\tilde{p}_{it}$, $\tilde{s}^{out}$ and $\tilde{s}_{kt}^{in}$ denote the fixed values of the corresponding variables, which are specialized in the following.

Concerning a power plant $i \in \mathcal{I}$, the corresponding constraints are always satisfied if the produced power p_{it} is set to the frozen value $\bar{p}_{it_r^*}$ for all $t \in \mathcal{T}_r \cup \mathcal{T}_r^{app}$. In particular, its upper and lower bounds are complied, as they are constant over the planning horizon. Also the minimum runtime and downtime restrictions are satisfied as the plant is neither switched on or off for $t \in \mathcal{T}_r \cup \mathcal{T}_r^{app}$.

For a storage $j \in \mathcal{J}$, we follow a different strategy. If $\bar{s}_{jt_r^*} = s_j^{max}$, we neither charge nor discharge and set $s_{jt} = \bar{s}_{jt_r^*}$ for all $t \in \mathcal{T}_r \cup \mathcal{T}_r^{app}$. Otherwise, the storage is charged at full power until $s_{j\tilde{t}} = s_j^{max}$ for a $\tilde{t} \in \mathcal{T}_r \cup \mathcal{T}_r^{app}$ and afterwards the storage level is kept at this level. If $|\mathcal{T}_r \cup \mathcal{T}_r^{app}|$ is large enough, this is always possible as $s_k^{in,min} = 0$ and consequently, the lower bound on the terminal storage level (3.9) is satisfied. On the other hand, if $|\mathcal{T}_r \cup \mathcal{T}_r^{app}|$ is not large enough and s_j^{max} is not reached within these time steps, we use (5.5) for time step t_r^*, which is

$$s_{jt_r^*} \geq \bar{s}_{jT} - (T - t_r^*) \sum_{k \in \mathcal{K}_j} f_k^{in,max} + \alpha_j^{in}(1 - y_{jt_r^*}^{in}),$$

where T denotes the number of time steps of the entire planning horizon. Hence, the condition (3.9) on the terminal storage is still satisfied, charging at full capacity in the last $T - t_r^*$ time steps. The remaining constraints associated with an energy storage are fulfilled by construction of this solution.

In conclusion, we are able to generate a feasible solution based on the fixation in t_r^*, following the strategy described above. □

Having shown that all subproblems P_r are feasible, altogether we obtain a feasible solution for the original problem by retrieving the values from the solutions of previous subproblems.

Besides the relaxation of the integrality conditions, we are also interested in the application of the approximation strategy S_1. Thus, we aim on transferring the result from the previous theorem to this variant. The use of approximation strategy S_1 affects the modeling of the approximated period more than the relaxation of the integrality restrictions, as not all integer feasible solutions are necessarily feasible if parts of the problems are approximated.

In order to guarantee feasibility of the subproblems P_r, the approximation of the piecewise linear functions $f_j^{in}(s_{kt}^{in})$, corresponding to the charging units $k \in \mathcal{K}_j$ of energy storages $j \in \mathcal{J}$, needs to be restricted. In accordance with assumption $s_k^{in,min} = 0$ of the previous theorem, we restrict the following studies to approximation $\bar{S}_1$, where $\eta_j^{in}(x)$ is approximated by a constant efficiency $\bar{\eta}_j^{in}$, i.e., $g(s_{kt}^{in}) = \bar{\eta}_j^{in} \cdot s_{kt}^{in}$. Thus, it is still possible to charge arbitrarily small amounts. Additionally, we require that $\bar{\eta}_j^{in}$ has to satisfy $\bar{\eta}_j^{in} s_k^{in,max} \geq f_k^{in,max}$ in order to prevent infeasibility caused by the approximated time steps. Note that if in the exact problem formulation, the storage levels can be chosen such that the final storage level can be reached, this condition ensures that using the approximated charging formulation the final storage level restriction can also be satisfied.

We remark that the approximation of the remaining efficiencies does not affect the feasibility of the subproblems, most of them appearing only in the objective function. Using Theorem 5.3, we obtain the following result for the approximation strategy $\bar{S}_1$:

Corollary 5.4 *Let the fixed charging efficiency $\bar{\eta}_j^{in}$ satisfy that $\bar{\eta}_j^{in} s_k^{in,max} \geq f_k^{in,max}$ for all $k \in \mathcal{K}_j$ with $j \in \mathcal{J}$ within the approximation strategy $\bar{S}_1$. Under the assumptions of Theorem 5.3, subproblem P_r is feasible in every iteration r if P is feasible.*

Proof. Let P_r denote the subproblem generated in iteration r and let t_r^* denote the last time step of the fixed period. For the proof, we create the auxiliary problem $\bar{P}_r$ which differs from subproblem P_r in the formulation of the approximated period $\mathcal{T}_r^{app}$ which is, in that case, formulated exactly as well. This means that $\bar{P}_r$ consists of a fixed period comprising all $t \leq t_r^*$ and an exact period containing all $t > t_r^*$. Using Theorem 5.3, we know that in iteration r all problems $\bar{P}_r$ are feasible if the storage level $s_{jt_r^*}$ satisfies condition (5.5). Thus, it remains to prove that if $\bar{P}_r$ is feasible, there also exists a feasible solution of P_r.

Again, we show the feasibility by construction of a feasible solution with respect to the fixations in time step t_r^*. Therefore, we follow the construction strategy described in the latter proof and adapt it to the problem formulation of $\bar{P}_r$ for the approximated period. As the approximations of the efficiency functions of power plants only appear in the objective function, the constraints of P_r and $\bar{P}_r$ only differ in the storage balance equation (3.29). Remember that for the construction of the feasible solution concerning all $t > t_r^*$, the storage is charged at full capacity until the upper bound of the storage level s_j^{max} is reached. Using that $\bar{\eta}_j^{in} s_k^{in,max} \geq f_k^{in,max}$ for all $k \in \mathcal{K}_j$ with $j \in \mathcal{J}$ together with constraint (5.5), we can always ensure that the terminal storage condition is satisfied. □

Finally, we consider the extension of the approximation strategy $\bar{S}_1$ to $\bar{S}_1^*$ as described above. As for $\bar{S}_1^*$ the last time steps of the approximated period are completely neglected, the resulting subproblem P_r is a relaxation of the subproblem $\bar{P}_r$, applying only $\bar{S}_1$. Consequently, all subproblems are feasible, yielding a feasible solution of the original problem at the end of the algorithm.

Altogether, we can only exclude the appearance of infeasible subproblems entirely if the parameters comply with conditions (A) and the approximation strategy is chosen appropriately. For instance, assume that $s_k^{in,min} > 0$ for the units $k \in \mathcal{K}_j$ of a storage $j \in \mathcal{J}$, it may occur that the final storage level can not be reached as a charging operation of at least $s_k^{in,min}$ in the last time step would exceed the maximum storage capacity.

Thus, aiming at the consideration of various test instances as well as on a flexible algorithm, the handling of infeasible subproblems has to be specified. As the infeasibility may result either from fixations or from the approximation strategy applied, we follow a two step approach. In particular, the following method is executed if an infeasible subproblem P_r occurs in iteration r in Step 4 of Algorithm 5.2:

1. The formulation of the approximated period $\mathcal{T}_r^{app}$ is changed to integrality relaxation R or R^*, respectively. Then problem P_r is resolved with adapted approximated period.

2. If P_r is still infeasible, a backward grouping step is performed as proposed by [BGGG06]. In detail we set

   $\mathcal{T}_{r-1} = \mathcal{T}_{r-1} \cup \mathcal{T}_r,$

   $\mathcal{T}_k = \mathcal{T}_{k+1}$ for all $k \in \{r, \ldots, R-1\},$

   $r = r - 1,$

   As a consequence, the number R of subproblems is decreased by one. If $r > 1$ execute Step 4 again using the restructured subdivision of the planning horizon. If $r = 1$, go to Step 2 and resolve P_1 with updated exact period.

Regarding the first step, infeasibilities based on approximations can be excluded. Step two provides the possibility of resetting variables already fixed. At worst, this approach may result in solving the complete exact problem in one iteration. But we remark that due to the additional storage bounds (5.5), the appearance of infeasible subproblems is very unlikely.

Summarizing, if this method is incorporated in the implementation of the algorithm, the rolling horizon heuristic terminates with a feasible solution, supposed the original problem is feasible. For computational results of the rolling horizon approach applied to the deterministic problem we refer to Section 8.3.1 and to [EMM+09].

5.3 An Approximate-and-Fix Approach to the *S-OPGen* Problem

For multistage stochastic programs less literature is known concerning primal heuristics applied to power generation problems, see also Section 2.4. With regard to the application of relax-and-fix strategies, the authors of [AAEO00] and [BGGG06] successfully applied this approach to multistage stochastic programs. In [AAEO00] a binary stochastic air traffic flow management problem is investigated. A basic version of relax-and-fix is implemented in order to provide good feasible solutions for the large scale deterministic equivalent. In [BGGG06] an enhanced version of relax-and-fix is applied to a stochastic lot sizing problem. They present different time partitioning policies and exploit the specific structure of the problem in order to prevent infeasibilities of the subproblems. The good results reported in these contributions motivate the application of this approach to the stochastic problem *S-OPGen*.

A further aspect encouraging the use of relax-and-fix is the good computational experiences of the rolling horizon heuristic for the deterministic case, see Section 8.3.1. Showing partly the same structure as the *D-OPGen* problem, i.e., considering one scenario, the adaptation of the deterministic version to the stochastic one seems promising. Therefore, the time partitioning strategy used in the rolling horizon approach is transferred to the scenario tree formulation as well as the investigation and the handling of infeasible subproblems. Additionally, the algorithm is also enhanced by the problem specific approximation strategies described above, yielding an *approximate-and-fix* heuristic. In the following, we present the adapted concept of approximate-and-fix, which is tailored to solve problem instances of *S-OPGen*. Details concerning problem-specific decisions like the choice of algorithmic parameters can also be found in the diploma thesis [Ric08].

We start by discussing how to construct subproblems P_r in iteration r under the consideration of the scenario tree formulation. Recall that so far, the problem was decomposed according to the subdivision of the planning

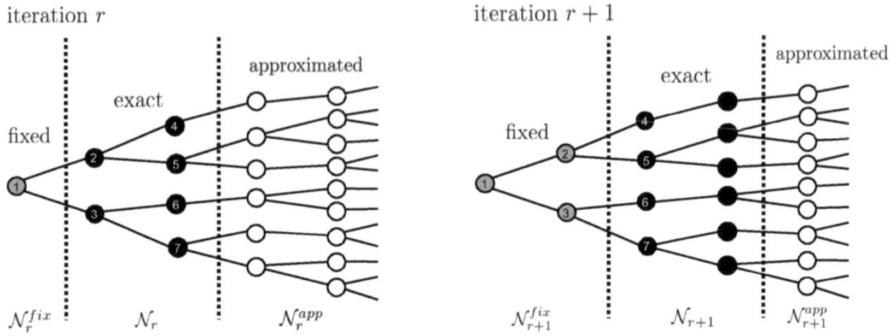

Figure 5.3: Subdivision of the scenario tree

horizon into several periods. As the problem structure of the deterministic case appears in the description of one scenario of the stochastic problem, we aim at directly transferring this approach to the stochastic problem. Hence, we want to group the variables in dependence to their time stage.

Therefore, the set of time steps $\mathcal{T}$ is classified into R subsets $\mathcal{T}_r$, according to the construction requirements (5.1) presented for the rolling horizon framework. Based on these subsets, the set of nodes $\mathcal{N}$ of the scenario tree Γ is subdivided into R subsets $\mathcal{N}_r$, i.e.,

$$\mathcal{N}_r = \{n \in \mathcal{N} \mid t(n) \in \mathcal{T}_r\}.$$

Remember that $t(n)$ denotes the time stage of node n in the scenario tree. Analogously, we define the sets

$$\begin{aligned} \mathcal{N}_r^{fix} &= \{n \in \mathcal{N} \mid t(n) < \underline{t}_r\}, \\ \mathcal{N}_r^{app} &= \{n \in \mathcal{N} \mid t(n) > \bar{t}_r\}, \end{aligned}$$

where $\underline{t}_r$ and $\bar{t}_r$ denote the first and the last time step of $\mathcal{T}_r$, respectively. The resulting tripartition, i.e., the fixed, the exact and the approximated part, is illustrated in Figure 5.3.

Note that in iteration r the number of integer variables depends on the number of nodes in $\mathcal{N}_r$. Naturally, the number of nodes per time stage increases with growing r and consequently, the number of integer variables to be considered. But as the variables corresponding to nodes $n \in \mathcal{N}_r^{fix}$ are fixed, subproblem P_r decomposes into several independent subproblems. In detail, the number of independent subproblems corresponds to the number of nodes associated with the first time stage of the exact period which is denoted by $\underline{t}_r$. These nodes are comprised in set $\mathcal{N}_{\underline{t}_r} = \{n \in \mathcal{N} \mid t(n) = \underline{t}_r\}$. Thus, $|\mathcal{N}_{\underline{t}_r}|$ subproblems Q_{rk} with $k \in \{1, \ldots, |\mathcal{N}_{\underline{t}_r}|\}$ have to be solved, where each subproblem Q_{rk} only involves a subset of the integer variables considered in iteration r.

In order to illustrate this approach exemplarily, we consider the scenario tree shown in Figure 5.3 with respect to iteration r. Having fixed the variables in the root node 1, two independent subproblems Q_{r1} and Q_{r2} can be formulated. Each of them corresponds to one of the subtrees, where the root node is node 2 or node 3, respectively. Passing to iteration $r+1$, the set $\mathcal{N}_{r+1}$ is created by transposing the considered time stages by one, i.e., $T^{shift} = 1$ and $T^{ex} = 2$, compare (5.1). Based on nodes $4, 5, 6$, and 7 we have to solve four independent subproblems in iteration $r+1$.

Now, we focus on the approximation strategies introduced in Section 5.2.1 and discuss their adaptation to the approximate-and-fix framework. Remember that strategy S_1 approximates the piecewise linear functions appearing in the description of energy storages and power plants. As only variables within one time step are affected by this approximation, the strategy can be transferred directly to the *S-OPGen* problem. In contrast, approximation S_2 yields a coarsening of the time partitioning, affecting the structure of the corresponding scenario tree which is also coarsened. Since the computational results using S_2 are not as promising as those obtained by S_1, see Section 8.3.1, strategy S_2 is not considered for the approximate-and-fix approach. The good performance of strategy S_1 results from the small impact on the structure of problem by reducing the number of binary variables drastically in contrast to S_2. Finally, the extended relaxation R^* and approximation S^* described on page 65 provide a bisection of the approximated part, by approximating the problem corresponding to nodes at the beginning of this period and completely neglecting the last part. Clearly, this approach is applicable to the problem *S-OPGen* using integrality relaxation R or S_1 for the approximated time steps. Hence, R^* and S_1^* are integrated in the implementation of the approximate-and-fix algorithm.

Finally, we address the feasibility of a subproblem P_r in iteration r. In [BGGG06] a related approach can be found, where the authors investigate the feasibility of subproblems applying relax-and-fix to a multistage stochastic lot-sizing problem. In contrast to our approach, they identify "representative" scenarios based on the specific structure of the lot-sizing problem in order to avoid the occurrence of infeasible subproblems. With respect to the *S-OPGen* problem, we develop an approach which is based on the results obtained from the deterministic case. Consequently, by requiring that

$$s_{jn} \geq \max\{s_j^{min}, \bar{s}_{Tj} - \frac{T - t(n)}{\tau} \sum_{k \in \mathcal{K}_j} f_k^{in,max} + \alpha_j^{in}(1 - y_{jn}^{in})\}, \qquad (5.6)$$

tightened bounds on the storage level s_{jn} of a storage $j \in \mathcal{J}$ in node $n \in \mathcal{N}$ are formulated for the stochastic problem. Here, $\bar{s}_{jT}$ describes the minimum terminal storage level of storage j, which is the same in all leaf nodes of the scenario tree.

Using inequalities (5.6) together with restrictions (A) on the parameter set of *S-OPGen*, we can transfer the feasibility results of Section 5.2.2 to the stochastic problem formulation. In particular, we start by considering integrality relaxation for the approximated period, yielding a generalization of Theorem 5.3. In the theorem and in the proof, the *S-OPGen* problem in abbreviated by P.

Theorem 5.5 *Let conditions (A) be fulfilled by the parameter set of P and assume inequalities (5.6) are added to the description of P. Using integrality relaxation for the approximated period, all subproblems P_r of the approximate-and-fix heuristic are feasible if P is feasible.*

Proof. This proof is done in analogy to the proof of Theorem 5.3. Thus, we restrict the following description to the basic differences resulting from the stochastic formulation.

Let t_r^* denote the last time step of the fixed period $\mathcal{T}_r^{fix}$ and let $\mathcal{N}_{t_r^*}$ contain all nodes of this time stage. In the following, we show that a subproblem P_r is feasible by constructing a feasible solution which is based on the fixed variables associated with nodes $n \in \mathcal{N}_{t_r^*}$.

For $r = 1$, problem P_1 provides a linear relaxation of the original problem and thus is feasible. Let $r > 1$. As all variables of the node $n \in \mathcal{N}_{t_r^*}$ are fixed, problem P_r decomposes into $|\mathcal{N}_{t_r^*+1}|$ independent subproblems Q_{rk} with $k \in \{1, \dots, |\mathcal{N}_{t_r^*+1}|\}$. Note that $\mathcal{N}_{t_r^*+1}$ contains all nodes corresponding to the first time step of the exact period. Considering a subproblem Q_{rk}, we construct a feasible setting of all variables associated with this problem in analogy to the construction described in the proof of Theorem 5.3. In particular, for each scenario of the corresponding subtree, the variables are set in accordance with this construction strategy, with the exception of the imported power x_n. This variable appears within the demand condition (3.1), which is the only constraint affected by the stochastic process.

As the scenarios of the corresponding subtree include different supplies of wind energy, the value of x_n may vary with different scenarios. But as the imported power x_n is unbounded, it can be set to a value large enough so that the constraint is always satisfied. In detail we set

$$\bar{x}_n = -\sum_{i\in\mathcal{I}} \bar{p}_{in} - \sum_{j\in\mathcal{J}} \sum_{l\in\mathcal{L}_j} \bar{s}^{out}_{ln} + \sum_{j\in\mathcal{J}} \sum_{k\in\mathcal{K}_j} \bar{s}^{in}_{kn} - \omega_n + \delta_{t(n)},$$

based on the demand condition described in Section 3.2.2.

As the constraints (5.6) are added to the problem formulation, the satisfaction of the lower bound on the terminal storage level (3.9) is also ensured. Thus, a feasible solution can be constructed for each subproblem Q_{rk} resulting in a feasible solution for problem P_r in iteration r. □

Based on Corollary 5.4, we extend the previous result to subproblems which are created using approximation strategy S_1. In order to be capable to charge arbitrarily small amounts by charging unit $k \in \mathcal{K}$ of a storage $j \in \mathcal{J}$, we consider the approximation strategy $\bar{S}_1$, which approximates the efficiency function $f(s^{in}_{kn})$ by the linear function $g(s^{in}_{kn}) = \bar{\eta}^{in}_j s^{in}_{kn}$ with constant efficiency $\bar{\eta}^{in}_j$.

Corollary 5.6 *Let the fixed charging efficiency $\bar{\eta}^{in}_j$ satisfy that $\bar{\eta}^{in}_j s^{in,max}_k \geq f^{in,max}_k$ for all $k \in \mathcal{K}_j$ with $j \in \mathcal{J}$ within the approximation strategy $\bar{S}_1$. Under the assumptions of Theorem 5.5, the subproblem P_r is feasible in each iteration r if P is feasible.*

Proof. This proof can be done in analogy to the proof of Corollary 5.4. □

Concerning the use of the extended relaxation R^* or approximation strategy $\bar{S}^*_1$, the subproblems remain feasible if P is feasible. Note that a subproblem P_r created by using strategy R^* or $\bar{S}^*_1$ provides a relaxation of subproblem $\tilde{P}_r$ obtained by applying only R or $\bar{S}_1$, respectively.

As for the rolling horizon heuristic, the possibility of infeasible subproblems can not be excluded entirely, when solving test instances which do not satisfy conditions (A). The strategies of handling infeasible subproblems can be transferred directly to the approximate-and-fix heuristic and we refer to Section 5.2.2.

In summary, most of the ideas developed for the deterministic case can be transferred straightforward to the stochastic one and altogether, the rolling horizon approach can be successfully adapted to *S-OPGen* problem yielding the approximate-and-fix heuristic described above. Thus, we have developed a flexible algorithm, which is able to create good feasible solutions for the stochastic problem in very good running times, see Sections 8.3.1 and 8.3.2. The tuning of the parameters is confined to the three values T^{ex}, T^{shift}, and T^{app}, which limits the effort to adjust the algorithm to different instances. In particular, good standard values of these parameters for instances of the *D-OPGen* and *S-OPGen* problems are suggested in Section 8.3.

Chapter 6

A Scenario Tree-Based Decomposition of Multistage Stochastic Mixed-Integer Problems

Based on our problem formulation of Section 3.2, we are interested in solving optimization problems where a set of parameters is uncertain. Modeling uncertainty via a set of scenarios and describing their relationship by the corresponding scenario tree, we obtain a multistage stochastic mixed-integer program (SMIP). As nonanticipativity constraints have to be respected, the deterministic problems associated with one scenario cannot be solved separately. Additionally, we want to consider problems where integer restrictions can appear in any stage of the problem, which may even make the solution of a one-scenario subproblem difficult. Furthermore, the size of problems normally grows very quickly with increasing number of time stages and scenarios considered in the model. In this chapter, we present a decomposition approach in order to solve the *S-OPGen* problem which shows the potential of solving a wide range of related problems.

The solution of multistage stochastic mixed-integer programs still poses a great challenge from the computational points of view as it comprises integer as well as stochastic aspects in one model. The need for modeling combinatorial decisions combined with uncertainty motivated a number of contributions, exemplarily presented in the sequel. One of the first papers concerning the solution of two-stage SMIPs was published by [LL93] proposing the Integer L-Shaped method. The algorithm is based on a branch-

and-cut procedure, where optimality cuts are generated for a fixed binary first stage solution. [CS99] propose a dual decomposition algorithm which applies to two-stage problems with integer variables in both stages employing Lagrangian relaxation for the decoupling. As stated by the authors, this method can also be used to solve multistage problems, see also [NR00]. The authors of [LS04] follow a branch-and-price approach to solve multistage SMIPS, using column generation to compute lower bounds. The algorithm described by [AAEG+03] and [AAEO03] relaxes nonanticipativity and integrality restrictions, yielding single scenario linear subproblems. A branch-and-bound procedure is used to restore feasibility, where each scenario has its own branch-and-bound tree. In [HS08] also a scenario decomposition is used, but integrality is maintained. By branching on nonanticipativity constraints, feasibility is reestablished.

Most of these contributions show some similarities to our method, as they also use a branch-and-bound approach in combination with a decomposition method. But rather than using a scenario decomposition, our approach is based on the subdivision of the corresponding scenario tree, which will be presented in the following sections. Finally, we refer to [RS01, Sch03], who provide an overview on further literature regarding modeling and solution approaches with respect to multistage stochastic programs with integer variables.

The remainder of the chapter is organized as follows. In Section 6.1, we start with the description of the basic idea of the proposed algorithm. Section 6.2 addresses the reformulation of the stochastic problem, which is based on the decomposition of the scenario tree into subtrees. In Section 6.3, we present the branch-and-bound method applied to the decomposed problem focusing on the general idea regarding the computation of lower bounds and branching. Finally, an extension of the branch-and-bound algorithm is developed, applying Lagrangian relaxation in order to generate tighter lower bounds in Section 6.4. Detailed information about the algorithmic implementation is given in Chapter 7.

6.1 Motivation and Idea

In this section, we describe the motivation for developing a new method in order to solve the *S-OPGen* as well as related problem and present the basic idea of the proposed algorithm. By formulating the stochastic problem as described in Section 3.2.2, we obtain a large-scale, block-structured mixed-integer optimization problem. Algorithmically, this structure makes the problems attractable for decomposition approaches. Indeed, the block-structure motivates the use of decomposition methods, as they provide the possibility of splitting problems of huge size into manageable subproblems. Especially in the linear case, successful decomposition approaches have been developed, see e.g. [BL97]. But also for problems including integrality restrictions, decomposition approaches are very promising as indicated above.

Currently, decomposition approaches for the solution of multistage mixed-integer programs are mainly based on scenario decomposition or relaxation of coupling constraints between different power units in case of power generation problems, as mentioned above. Nevertheless, the former approach cannot be used directly for the solution of the *S-OPGen* problem, as even the solution of a one scenario-problem can be computationally challenging, see Section 8.3.1. The latter approach is successfully applied to problems whose one-unit subproblems can be solved efficiently. For instance, in [NR00], the subproblems can be restated as combinatorial multistage stochastic programs, which are solved by stochastic dynamic programming. However, the *S-OPGen* shows different problem characteristics, due to the detailed modeling of the facilities.

Hence, for the solution of the *S-OPGen* problem, we have developed a new decomposition method which is mainly motivated by the following two observations. First, the problem shows a loose connectivity with respect to variables associated with different nodes of the scenario tree. In particular, two time steps are only coupled by the storage balance equation, the minimum run time and down time restriction and the upper bound on the power gradient, see Section 3.1.3. The second and more important observation is based on the computational investigation in the course of the approximate-and-fix heuristic, presented in Section 8.3.2. Namely, the fixation of variables at a selected node of the scenario tree has only little impact on the optimal solution values of variables associated with nodes which are sufficiently far away. This means that in most cases, the optimal decisions corresponding to a node n are not changed if a variable of a further node m is fixed and the distance of n and m exceeds a certain path length in the scenario tree. Consequently, the goal is to employ a decomposition which generates subproblems whose coupling between each other is hardly correlated and thus, exploits the lack of sensibility described above.

The basic concept of the developed algorithm includes this decomposition suggestion and combines it with a branch-and-bound procedure. The underlying idea is based on the partition of the scenario tree into several smaller subtrees by defining so called *split nodes*, where the tree is split up.

Based on this subdivision, the resulting subproblems are formulated independently. Note that in contrast to the scenario decomposition only variables corresponding to split nodes need to be doubled. The formulations are connected by adding so called *coupling constraints* yielding a reformulation of the original problem. If the coupling constraints are relaxed, the problem decouples into a collection of separate subproblems, which can be solved independently providing a lower bound on the optimal objective function value. In order to ensure feasibility, the decomposition is embedded within a branch-and-bound framework. This means that by branching on pairs of variables the satisfaction of the coupling constraints is restored.

The decomposition based branch-and-bound approach provides the following advantages. Decomposing the problem with respect to predefined split nodes allows us to determine the size of the subproblems depending on the individual problem. Here, we remark that the resulting subproblems are still mixed-integer formulations which makes a suitable size desirable in order to achieve a good performance. In Section 6.3, we will show that in each branch-and-bound node at most one subproblem has to be solved in order to obtain a lower bound on the optimal function value. Indeed, subproblems with identical branching bounds may appear various times during the solution process. This fact can be exploited by a suitable caching procedure which stores already solved subproblems, see Section 7.3.2. Moreover, we can benefit from flexibility of the branch-and-bound approach concerning the application of further techniques to speed up the algorithm, such as the integration of Lagrangian relaxation, problem specific heuristics, branching strategies and separation algorithms.

In summary, the notable size of the *S-OPGen* problem instances and the good experience to decompose this type of problems encourage the development of the decomposition based branch-and-bound approach. This motivation is stressed by the additional advantages provided by the chosen decomposition approach as described above. In particular, it allows the exploitation of the loose connectivity between the time steps as well as problem specific characteristics of the *S-OPGen* problem. In the following, the developed method is presented in detail which aims at solving the linearized *S-OPGen* problem to global optimality.

6.2 Reformulation and Decomposition of the Stochastic Problem

In the sequel, a reformulation of the stochastic problem is described which provides the basis for the decomposition. In accordance with the definition of a multistage stochastic problem given Section 3.2, the original problem called *OP* can be described as follows:

$$(OP)\quad \begin{aligned} \min \sum_{n\in\mathcal{N}} \pi_n c_n^\top x_n & & & \\ \text{s.t.} \qquad W_1 x_1 &= b_1, & & \\ T_n x_{p(n)} + W_n x_n &= b_n, & &\text{for } n \in \mathcal{N}\backslash\{n_1\}, \\ x_n &\in X_n, & &\text{for } n \in \mathcal{N}. \end{aligned} \tag{6.1}$$

Remember that $\mathcal{N}$ comprises all nodes of the corresponding scenario tree rooted at node n_1 and parameter π_n reflects the probability of node n. The index $p(n)$ refers to the predecessor of node n in the tree, and the vector x_n comprises all variables belonging to node n. The set X_n is defined by integrality restrictions as well as lower and upper bounds on the variables of node $\tilde{n}$. The resulting block structure of the matrix is shown exemplarily in Figure 6.1, where the problem formulation is based on the six-node scenario tree visualized on the left hand side of the figure. More precisely, the k-th column of blocks with $k \in \{1, \dots, 6\}$ is associated with the variables of node n_k.

The main idea of decomposing the problem into smaller subproblems is based on the consideration of the scenario tree $\Gamma = (\mathcal{N}, \mathcal{E})$. At the beginning, we choose a subset $\mathcal{N}_{split}$ of $\mathcal{N}\backslash\{n_1\}$, which contains all nodes where the tree is split up. Hence, they are called *split nodes*. For each split node $n \in \mathcal{N}_{split}$ a new node $\tilde{n}$ is created, where n and $\tilde{n}$ are not connected. In-

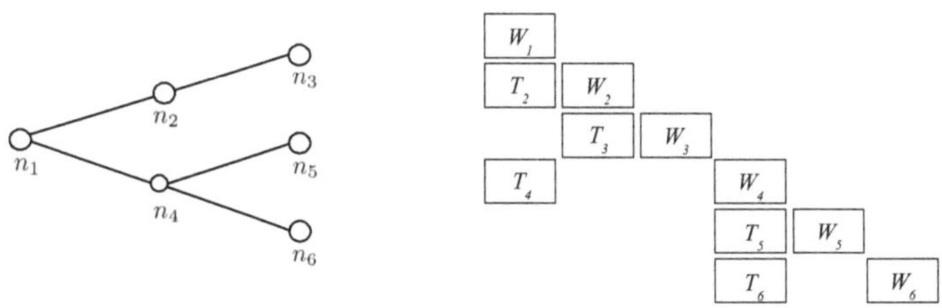

Figure 6.1: A block-structured matrix with corresponding scenario tree

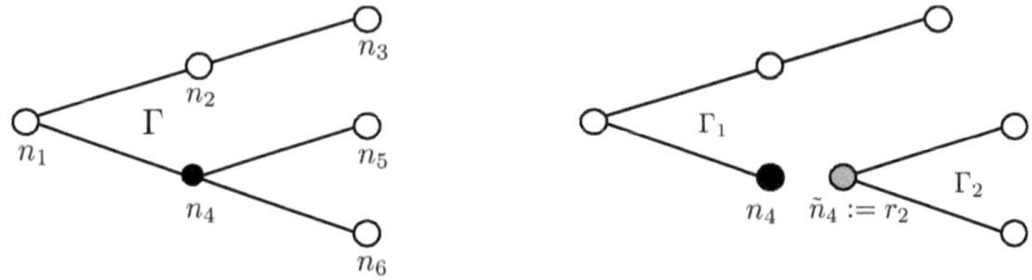

Figure 6.2: Exemplary splitting of a scenario tree with 6 nodes

stead, changing all edges (n, m) with m being a direct successor of n to an edge $(\tilde{n}, m)$, a new subtree is created, for which $\tilde{n}$ forms the root node. The resulting subtrees are denoted by $\Gamma_k = (\mathcal{N}_k, \mathcal{E}_k)$ with $k \in \mathcal{K} = \{1, \ldots, K\}$ for some $K \in \mathbb{N}$ and we refer to the set of nodes of tree Γ_k as $\mathcal{N}_k$. The procedure is visualized in Figure 6.2, where the split node n_4 is marked in black and the duplicated node $\tilde{n}_4$ in gray yielding two subtrees Γ_1 and Γ_2.

Returning to the problem formulation, we now discuss the treatment of a variable x_{ni} assigned to a split node $n \in \mathcal{N}_{split}$ and $i \in \mathcal{J}_n$, where $\mathcal{J}_n$ describes the index set of vector x_n. When creating a new node $\tilde{n}$ of a split node n, only some variables of node n need to be doubled, as not all variables of this node impact the subproblem corresponding to the subtree with root node $\tilde{n}$. More precisely, this concerns all variables x_{ni} which connect any successor node m with node n in any constraint of the original problem P. In the following, we denote the set of indices of the time-connecting variables corresponding to node n by $\mathcal{I}_n \subseteq \mathcal{J}_n$. Thus, for each x_{ni} with $i \in \mathcal{I}_n$ and $n \in \mathcal{N}_{split}$ a new variable $x_{\tilde{n}i}$ is created and only these variables are assigned to node $\tilde{n}$. Based on the splitting of the scenario tree and the doubling of variables, the constraints are adapted correspondingly yielding K separable blocks of constraints. By weighting all $x_{\tilde{n}i}$ with zero in the objective function, the problem can be split into K separate subproblems Q_k with $k \in \mathcal{K}$. Consequently, the objective function of the subproblems consists of the summands of the original function concerning the nodes of the corresponding subtree Γ_k.

In order to describe the subproblems explicitly, we introduce the following notation. By

$$\mathcal{X}_k = \{(x_n)_{n \in \mathcal{N}_k} \quad | \quad T_n x_{p(n)} + W_n x_n = b_n, \quad \text{for all } n \in \mathcal{N}_k \backslash \{r_k\}, \qquad (6.2)$$
$$x_n \in X_n, \qquad \text{for all } n \in \mathcal{N}_k \ \},$$

we refer to the set of feasible solutions of subproblem $k \in \mathcal{K} \backslash \{1\}$. The new set X_{r_k} for the root node r_k of tree Γ_k is defined by integrality restrictions as well as lower and upper bounds on variables of node $\tilde{n}$ in analogy to the definition of X_n for $n \in \mathcal{N}$. For the definition of the set of feasible points $\mathcal{X}_1$ of the first subproblem, the system $W_1 x_1 = b_1$ has to be added to the constraints described above. Additionally, we denote by

$$\boldsymbol{x}_k = (x_n)_{n \in \mathcal{N}_k}$$

the vector of all variables associated with nodes $n \in \mathcal{N}_k$. With regard to the objective function, we comprise all terms corresponding to nodes $n \in \mathcal{N}_k \setminus \{r_k\}$ in

$$z_k(\boldsymbol{x}_k) = \sum_{n \in \mathcal{N}_k \setminus \{r_k\}} \pi_n c_n^\top x_n,$$

Altogether, a subproblem Q_k corresponding to subtree Γ_k for $k \in \mathcal{K}$ can be formulated as follows:

$$(Q_k) \qquad \begin{aligned} \min \quad & z_k(\boldsymbol{x}_k) \\ \text{s.t.} \quad & \boldsymbol{x}_k \in \mathcal{X}_k. \end{aligned} \tag{6.3}$$

Aiming at a reformulation of the original problem described in (6.1), the correct coupling between the subproblems must be ensured. Thus, for each split node $n \in \mathcal{N}_{split}$ and $i \in \mathcal{I}_n$, we introduce the following set of equations called *coupling constraints*:

$$x_{ni} = x_{\tilde{n}i}. \tag{6.4}$$

Thus, problem (6.1) can be reformulated as:

$$(P) \qquad \begin{aligned} \min \quad & \sum_{k \in \mathcal{K}} z_k(\boldsymbol{x}_k) \\ \text{s.t.} \quad & x_{ni} = x_{\tilde{n}i}, && \text{for } n \in \mathcal{N}_{split} \text{ and } i \in \mathcal{I}_n, \\ & \boldsymbol{x}_k \in \mathcal{X}_k, && \text{for } k \in \mathcal{K}. \end{aligned} \tag{6.5}$$

Finally, we introduce the function $\kappa : \bigcup_{k \in \mathcal{K}} \mathcal{N}_k \mapsto \mathcal{K}$ which maps a node n to the index $\kappa(n)$ of the corresponding subtree $\Gamma_{\kappa(k)} = (\mathcal{N}_{\kappa(k)}, \mathcal{E}_{\kappa(k)})$, i.e., $n \in \mathcal{N}_{\kappa(n)}$. In other words, $\kappa(n)$ indicates the subtree to which a node belongs. This function is used within the description of the branch-and-bound process described subsequently.

We remark that this splitting procedure is designed for problem formulations, where the constraints couple variables of at most two consecutive nodes. But note that those constraints which connect more than two consecutive nodes can be reformulated as constraints which couple only two consecutive nodes by introducing auxiliary variables. This means that this approach is applicable to a wider range of problems.

6.3 A Scenario Tree-Based Decomposition Combined with Branch-and-Bound

This section covers the general concept of the decomposition based branch-and-bound algorithm in order to solve the *S-OPGen* problem, whereas the algorithmic implementation is described in Chapter 7. The basic idea of this approach is to combine the presented decomposition with a branch-and-bound framework in order to generate an optimal feasible solution of the original problem. The latter approach is the most widely applied algorithm for solving mixed-integer optimization problems and basically consists of the following two phases: Firstly, the partition of the feasible region into smaller subsets called *branching* and secondly the *bounding*, which refers to the computation of lower and upper bounds on the optimal function value, see e.g. [NW88]. By performing a branching step the feasible set is partitioned recursively which can be represented by a *branch-and-bound tree*. The bounding step enables to disregard certain subproblems from search, called *pruning*, in order to avoid complete enumeration.

In the following, we specify the developed solution approach for multistage mixed-integer programs which couples a branch-and-bound method with the decomposition resulting from the subdivision of the corresponding scenario tree as described in Section 6.2. A summary of the entire procedure is given in Algorithm 6.1. It consists of six steps, which are *initialization, termination, problem selection, pruning, heuristic, and branching*. Subsequently, all steps are explained in detail.

As seen in Section 6.2, the multistage stochastic problem can be reformulated based on the definition of split nodes $\mathcal{N}_{split}$ of the corresponding scenario tree. The first step of the algorithm is the consideration of the relaxed problem $\bar{P}$ obtained by omitting the coupling conditions (6.4) from the reformulated problem P described in (6.5) which yields

$$(\bar{P}) \qquad \begin{aligned} \min \quad & \sum_{k \in \mathcal{K}} z_k(\boldsymbol{x}_k) \\ \text{s.t.} \quad & \boldsymbol{x}_k \in \mathcal{X}_k, \qquad \text{for } k \in \mathcal{K}. \end{aligned} \tag{6.6}$$

By construction, this relaxation enables us to solve the smaller subproblems Q_k separately. In order to obtain a first lower bound on the optimal objective function value of the original problem, each Q_k with $k \in \mathcal{K}$ is solved to optimality and the function values are summed up. We call this phase *initialization*, which is described in Step 1 of the Algorithm 6.1.

Algorithm 6.1: *Scenario Tree-Based Decomposition Combined with a Branch-and-Bound Algorithm (SD-BB)*

Input: A problem P as defined in (6.5)

Output: An optimal solution x^* of P with respect to accuracy δ or status "infeasible"

Step 1: Initialization
Initialize accuracy δ
Let $\mathcal{L}$ be the list of unsolved problems which is initialized by P
Consider the relaxation $\bar{P}$ by omitting the coupling conditions according to (6.6) and solve the resulting subproblems Q_1 to Q_K
Set the global upper bound $UB = \infty$ and **goto** Step 4

Step 2: Termination
If $\mathcal{L}$ is empty and $LB < \infty$ **then**
  Return optimal solution x^* if it exists and otherwise status "infeasible"

Step 3: Problem Selection and Relaxation
Choose a problem $P \in \mathcal{L}$ and update $\mathcal{L} = \mathcal{L} \backslash \{P\}$
Consider the relaxation $\bar{P}$ according to (6.6) and solve the affected subproblem Q_k marked in Step 6

Step 4: Pruning
If $\bar{P}$ is infeasible **then goto** Step 2
Else
  Let $\hat{\boldsymbol{x}}_{\bar{P}} = (\hat{\boldsymbol{x}}_1, \ldots, \hat{\boldsymbol{x}}_K)$ be an optimal solution of $\bar{P}$ and let $LB_{\bar{P}} = \sum_{k \in \mathcal{K}} z_k(\hat{\boldsymbol{x}}_k)$ be its objective function value
  If $LB_{\bar{P}} \geq UB$ **then** Prune and **goto** Step 2
  If $\hat{\boldsymbol{x}}_{\bar{P}}$ is feasible with respect to accuracy δ **then**
    If $LB_{\bar{P}} < UB$ **then**
      Update $UB = LB_{\bar{P}}$ and set best feasible solution $\boldsymbol{x}^* = \hat{\boldsymbol{x}}_{\bar{P}}$
      Delete all problems P' from $\mathcal{L}$ with $LB_{\bar{P}'} \geq UB$
    Prune and **goto** Step 2

Step 5: Heuristic
Based on $\hat{\boldsymbol{x}}_{\bar{P}}$, generate a feasible solution $\bar{\boldsymbol{x}}_P$ for P, applying a heuristic algorithm
Let $\bar{z}_P$ denote the corresponding objective function value
If $\bar{z}_P < UB$ **then**
  Delete all problems P' from $\mathcal{L}$ with $LB_{\bar{P}'} \geq \bar{z}_P$
  Set $UB = \bar{z}_P$ and $\boldsymbol{x}^* = \bar{\boldsymbol{x}}_P$

Step 6: Branching

Select a violated coupling constraint with corresponding variables x_{ni} and $x_{\tilde{n}i}$

Create P^+ and P^- according to (6.7) to (6.10)

Depending on the solution $\hat{\boldsymbol{x}}_{\bar{P}}$, mark the affected subproblem $Q_{\kappa(n)}$ in P^+ and $Q_{\kappa(\tilde{n})}$ in P^- or vice versa

Add P^+ and P^- to $\mathcal{L}$ and **goto** Step 2

Note that if any of the subproblems is infeasible, P is also infeasible by definition of the relaxation and thus the algorithm stops. Otherwise, let $\hat{\boldsymbol{x}}_k$ denote the optimal solution of subproblem Q_k. If $(\hat{\boldsymbol{x}}_1, \ldots, \hat{\boldsymbol{x}}_K)$ already satisfies the coupling conditions, then the solution is feasible for the original problem P and thus, optimal and the algorithms stops, too.

If none of these two trivial cases occur, a suitable *heuristic* is applied for the determination of an upper bound (UB) on the optimal function value which is addressed in Step 5. Either we may use a construction heuristic, which constructs a feasible solution from scratch, or we exploit the solution received in course of the computation of the lower bound in order to generate a feasible solution. On this aspect we focus in Section 7.4.

Subsequent to the heuristic, a *branching* step is performed, constituting the core of the algorithm, compare Step 6. By branching we aim at increasing the lower bound on the optimal function value as well as ensuring the relaxed coupling constraints again. Assuming that the solution $(\hat{\boldsymbol{x}}_1, \ldots, \hat{\boldsymbol{x}}_K)$ of the relaxation is infeasible for the original problem P, there exists a split node $n \in \mathcal{N}_{split}$ and index $i \in \mathcal{I}_n$ such that the corresponding coupling condition $x_{ni} = x_{\tilde{n}i}$ is violated. Recall that the subproblems are solved to optimality yielding integer feasible solutions.

Within the branching procedure, we distinguish the two cases, where either binary or continuous variables are involved in the violated coupling constraint. We start our considerations by assuming that x_{ni} and $x_{\tilde{n}i}$ are binary, violating the coupling constraints, i.e., $\hat{x}_{ni} = 0$ and $\hat{x}_{\tilde{n}i} = 1$ or vice versa. Then, the set of feasible solutions is subdivided into two parts by requiring that either

$$x_{ni} \leq 0 \quad \text{and} \quad x_{\tilde{n}i} \leq 0 \tag{6.7}$$

or

$$x_{ni} \geq 1 \quad \text{and} \quad x_{\tilde{n}i} \geq 1. \tag{6.8}$$

Note that in contrast to the commonly used branching on one variable here the problem is split up by adding two inequalities. Based on this branching step, two new subproblems are build in the branch-and-bound process, where the so called *right subproblem* $\bar{P}^+$ is obtained by adding inequalities (6.7) to the current problem $\bar{P}$ and the *left subproblem* $\bar{P}^-$ results from the addition of inequalities (6.8). Both problems are added to the list $\mathcal{L}$ of open problems.

As an important property, this branching procedure does not interfere with the separability of the decomposed formulation. To be more precise, let $Q_{\kappa(n)}$ denote the subproblem including variable x_{ni} and $Q_{\kappa(\tilde{n})}$ the subproblem including $x_{\tilde{n}i}$ as defined in (6.3). Recall that $\kappa(n)$ indicates the subtree to which node n belongs. Then, $\bar{P}^+$ is created by adding $x_{ni} = 0$ to $Q_{\kappa(n)}$ and $x_{\tilde{n}i} = 0$ to $Q_{\kappa(\tilde{n})}$. Analogously, $\bar{P}^-$ is obtained by adding $x_{ni} = 1$ to $Q_{\kappa(n)}$ and $x_{\tilde{n}i} = 1$ to $Q_{\kappa(\tilde{n})}$. Indeed, all subproblems Q_k with $k \in \mathcal{K} \backslash \{\kappa(n), \kappa(\tilde{n})\}$ are not affected by this branching step and altogether, the separability of the subproblems Q_k for all $k \in \mathcal{K}$ is maintained.

Concerning the branching on continuous variables, we face the problem of running into an infinite partitioning of the corresponding interval, as the branching points are not finite as in case of binary variables. In order to avoid infinite branching, we say that a coupling condition of a continuous variable is satisfied if $|x_{ni} - x_{\tilde{n}i}| \leq \delta$ for a fixed accuracy $\delta > 0$. On this basis we assume that the coupling condition of the continuous variables x_{ni} and $x_{\tilde{n}i}$ is violated by the current solution if $|\hat{x}_{ni} - \hat{x}_{\tilde{n}i}| > \delta$. Then, a branching point $b \in \mathbb{R}$ with

$$\min(\hat{x}_{ni}, \hat{x}_{\tilde{n}i}) < b < \max(\hat{x}_{ni}, \hat{x}_{\tilde{n}i})$$

is selected. The choice of an adequate branching point is discussed in Section 7.2.2. Using b, the feasible domain of the current problem is subdivided into two subdomains by requiring that either

$$x_{ni} \leq b \quad \text{and} \quad x_{\tilde{n}i} \leq b \tag{6.9}$$

or

$$x_{ni} \geq b \quad \text{and} \quad x_{\tilde{n}i} \geq b. \tag{6.10}$$

The creation of the subproblems $\bar{P}^+$ and $\bar{P}^-$ works analogously to the binary case. Clearly, the separability of the decomposed problems can also be maintained for the continuous variables.

A key point of the algorithm is that in each node of the branch-and-bound tree at most one subproblem Q_k has to be solved in order to compute a lower bound LB, which is explained in the following. As mentioned above, a branching step on the pair of variables $(x_{ni}, x_{\tilde{n}i})$ only affects the subproblems $Q_{\kappa(n)}$ and $Q_{\kappa(\tilde{n})}$ and the remaining subproblems Q_k with $k \in \mathcal{K}\backslash\{\kappa(n), \kappa(\tilde{n})\}$ are not changed. Looking closely at the subproblems $\bar{P}^+$ and $\bar{P}^-$, in both cases already one of the two added inequalities is already satisfied by the current solution $\hat{x}_{ni}$ and $\hat{x}_{\tilde{n}i}$. For an illustration, consider the binary variables x_{ni} and $x_{\tilde{n}i}$ in a branch-and-bound node which are chosen such that $\hat{x}_{ni} \neq \hat{x}_{\tilde{n}i}$. Regarding branching inequalities (6.7), either $\hat{x}_{ni}$ satisfies $x_{ni} \leq 0$ or $\bar{x}_{\tilde{n}i}$ satisfies $x_{\tilde{n}i} \leq 0$, respectively. Consequently, only $Q_{\kappa(n)}$ or $Q_{\kappa(\tilde{n})}$ needs to be solved in order to obtain the optimal value of subproblem $\bar{P}^+$. Remember that the optimal function value is computed by the sum of the values of the subproblems. The same holds for subproblem $\bar{P}^-$.

In summary, the branching strategy has the following three characteristics: Firstly, the branching is done on a pair of variables instead of only a single one and secondly, the separability of the subproblems Q_k is conserved, as the additional condition is separable as well. Finally, only two subproblems are affected by the branching step and at most one subproblem has to be solved in each branch-and-bound node for the determination of a new lower bound.

The algorithms *terminates* when all nodes of the branch-and-bound tree are processed or a certain threshold between the value of the best solution UB and the value of the smallest lower bound $LB_{\bar{P}}$ of all open subproblems is met. In detail, we set a relative tolerance ε on the gap which is computed by $\frac{|LB-UB|}{|UB|}$.

Otherwise an open problem P of the list $\mathcal{L}$ is *selected* and *relaxed* according to (6.6), i.e., the coupling constraints are omitted. As explained within the branching step, only one subproblem Q_k is affected by applying the branching procedure. Hence, by resolving Q_k an updated lower bound is computed. Details concerning the implementation of this step are described in Section 7.3.

Finally, we consider the *pruning* step, where we differentiate between the following three classical cases, see Step 4 of the algorithm: First, a node in the branch-and-bound tree is pruned if the corresponding problem $\bar{P}$ is infeasible, which means the affected subproblem Q_k need to be infeasible. Secondly, if the optimal objective function value of the relaxed problem $\bar{P}$

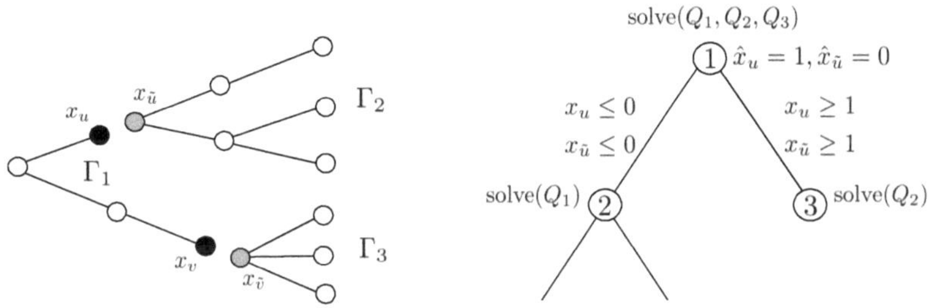

Figure 6.3: Decomposed scenario tree and corresponding branch-and-bound tree

exceeds the best upper bound found so far, the node is pruned as well. Finally, a node is pruned, if the optimal solution satisfies all coupling constraints with respect to the accuracy δ, as explained above.

In the following, we refer to the decomposition based branch-and-bound approach as the *SD-BB algorithm*. In order to clarify the procedure defined above, the following example is given.

Example 6.1 As illustrated in Figure 6.3, we consider a scenario tree which is decomposed into three subtrees denoted by Γ_1, Γ_2, and Γ_3, i.e., $K = 3$. Here, $\mathcal{N}_{split} = \{u, v\}$ implies the duplication of split node u to node $\tilde{u}$ with corresponding binary variables x_u and $x_{\tilde{u}}$. Analogously, node v and the duplicated node $\tilde{v}$ are associated with the binary variables x_v and $x_{\tilde{v}}$, respectively. Remember that x has to be time-connecting in order to be considered within the splitting procedure. Based on this subdivision, the subproblems Q_1, Q_2, and Q_3 are created as described in (6.3).
Solving the three subproblems independently, we receive a lower bound for the optimal function value by summing up $z_1(\hat{\boldsymbol{x}}_1)$, $z_2(\hat{\boldsymbol{x}}_2)$, and $z_3(\hat{\boldsymbol{x}}_3)$, where $\hat{\boldsymbol{x}}_k$ denotes the optimal solution of subproblem Q_k with $k \in \{1, 2, 3\}$. In this example, the optimal solutions of the subproblems violate the coupling condition $x_u - x_{\tilde{u}} = 0$, taking the values $\hat{x}_u = 1$ and $\hat{x}_{\tilde{u}} = 0$. As shown in the branch-and-bound tree on the right hand side of Figure 6.3, we branch on the condition $x_u \leq 0$ and $x_{\tilde{u}} \leq 0$ as well as on $x_u \geq 1$ and $x_{\tilde{u}} \geq 1$.
Regarding node 2 of the branch-and-bound tree, inequality $x_u \leq 0$ is added to Q_1 and $x_{\tilde{u}} \leq 0$ to Q_2, respectively, yielding the new problem P^+. In order to compute a lower bound in node 2 only Q_1 need to be resolved as the optimal solution of Q_2 obtained in node 1 already satisfies this new restriction. In order to exploit this observation when problem P^+ is processed, subproblem Q_1 is marked in P^+.

6.4 Improving *SD-BB* by Applying Lagrangian Relaxation

In the sequel, we present a Lagrangian relaxation approach for the *S-OPGen* problem with the goal of generating tight lower bounds for the optimal function value of the problem. As this approach has an impact on the formulation of the subproblems used in the *SD-BB* framework, it is applied only in the root node of the branch-and-bound algorithm. Why it is reasonable to fix the Lagrangian multipliers during the branch-and-bound process is justified later in this section. The focus of this section is on the application of Lagrangian relaxation to our problem formulation as well as on its integration into our branch-and-bound algorithm. For a detailed description of Lagrangian relaxation we refer to [Geo74] and [NW88].

Lagrangian relaxation is currently a popular method for solving energy generation problems as it makes use of the separability of the problems. For unit commitment problems, this approach provides the advantage of decomposing the entire problem into smaller one-unit subproblems, which often can be solved with adapted efficient algorithms, see e.g. [NR00] and Section 2.4. Utilizing Lagrangian relaxation, the decision of which constraints are relaxed has a major influence on the quality of the algorithm. On the one hand, it is desirable to relax as much complicating constraints as possible as the resulting relaxed problem has to be resolved several times during the solution process. On the other hand, the relaxation of more constraints results in worse lower bounds of the objective function value. As explained above, we follow the approach of relaxing only few constraints associated with selected nodes. Although we must deal with the consequence of solving mixed-integer subproblems, they are of small size and can be solved within short running times.

6.4.1 Lagrangian Relaxation of Coupling Constraints

So far, we obtain a relaxation of the *S-OPGen* problem by completely neglecting the coupling constraints (6.4), which are

$$x_{ni} = x_{\tilde{n}i},$$

for all $n \in \mathcal{N}_{split}$ and $i \in \mathcal{I}_n$ yielding the subproblems Q_k with $k \in \mathcal{K}$ defined in (6.3). In order to receive tighter bounds, we introduce a Lagrangian multiplier vector $\boldsymbol{\lambda}_n \in \mathbb{R}^{|\mathcal{I}_n|}$ for each system of coupling constraints associated with a split node $n \in \mathcal{N}_{split}$. In short, we define $\boldsymbol{\lambda} = (\boldsymbol{\lambda}_n)_{n \in \mathcal{N}_{split}}$. By introducing the product of each coupling constraint with the corresponding Lagrangian multiplier to the objective function of the reformulated relaxed problem $\bar{P}$, we receive the Lagrangian dual function:

$$d(\boldsymbol{\lambda}) = \min_{\boldsymbol{x} \in \mathcal{X}} \sum_{k \in \mathcal{K}} z_k(\boldsymbol{x}_k) + \sum_{n \in \mathcal{N}_{split}} \sum_{i \in \mathcal{I}_n} \lambda_{ni}(x_{ni} - x_{\tilde{n}i}), \tag{6.11}$$

where $\boldsymbol{x} = (\boldsymbol{x}_k)_{k \in \mathcal{K}}$ and $\mathcal{X} = \mathcal{X}_1 \times \ldots, \times \mathcal{X}_K$. Note that for variable multiplier vectors $\boldsymbol{\lambda}_n$, the dual function is not decomposable anymore as the multipliers couple the subproblems in the objective function. Nevertheless, we want to rewrite the function $d(\boldsymbol{\lambda})$ as the sum of K functions $d_k(\boldsymbol{\lambda})$ such that each $d_k(\boldsymbol{\lambda})$ only contains terms corresponding to nodes of subtree Γ_k. In order to formulate the functions explicitly, we define the set

$$\mathcal{N}_{split}^k = \mathcal{N}_{split} \cap \mathcal{N}_k$$

comprising all split nodes of subtree k. Thus, we are able to formulate $d_k(\boldsymbol{\lambda})$ by

$$d_k(\boldsymbol{\lambda}) = \min_{\boldsymbol{x}_k \in \mathcal{X}_k} z_k(x_k) + \sum_{n \in \mathcal{N}_{split}^k} \sum_{i \in \mathcal{I}_n} \lambda_{ni}\, x_{ni} - \sum_{i \in \mathcal{I}_{r_k}} \lambda_{r_k,i}\, x_{r_k,i} \tag{6.12}$$

for all $k \in \mathcal{K} \backslash \{1\}$. The first sum comprises all products of a Lagrangian multiplier with a variable which corresponds to a split node of Γ_k. As the root node r_k is not an element of $\mathcal{N}_{split}^k$, all products corresponding to r_k are treated separately in the last sum. For $k = 1$, the last sum vanishes, as the root node $\tilde{n}_1$ is not a duplication of a split node. It is a well known result that for any $\boldsymbol{\lambda}$ the value of $d(\boldsymbol{\lambda})$ provides a lower bound on the optimal value of the original problem P, see e.g. [NW88]. As we aim at maximizing the lower bound, we consider the dual problem

$$\max_{\boldsymbol{\lambda} \in \mathbb{R}^p} d(\boldsymbol{\lambda}), \tag{6.13}$$

where $p = \sum_{n \in \mathcal{N}_{split}} |\mathcal{I}_n|$ represents the number of all coupling constraints. Consequently, we are interested in determining good values for the Lagrangian multipliers in order to receive a large value of $d(\boldsymbol{\lambda})$. Knowing that the function $d(\boldsymbol{\lambda})$ is piecewise linear and concave, see e.g. [NW88], a common approach is to utilize a subgradient method for the maximization. Details on the algorithmic implementation of the approach are given in Section 7.3.1.

Clearly, when fixing the Lagrangian multipliers $\boldsymbol{\lambda}$ to $\bar{\boldsymbol{\lambda}}$, the separability of the dual problem with respect to the subtrees is restored, yielding the independent functions $d_k(\bar{\boldsymbol{\lambda}})$. This fact provides the basis for applying Lagrangian relaxation in the *SD-BB* algorithm and is exploited in the sequel.

6.4.2 Integration of Lagrangian Relaxation into the *SD-BB* Algorithm

Using the Lagrangian relaxation in the branch-and-bound framework has an impact on the entire solution process. Thus, we aim at integrating this approach into the branch-and-bound algorithm without destroying its basic properties as described in 6.3. Remember that as a key point of the algorithm at most one subproblem Q_k with $k \in \mathcal{K}$ needs to be solved in each branch-and-bound node in order to compute a lower bound. The procedure requires that the remaining problems Q_l with $l \in \mathcal{K} \setminus \{k\}$ are not affected in the current node. This property is maintained if the Lagrangian multipliers $\boldsymbol{\lambda}$ are constant during the solution process as explained above. Consequently, the determination of good values for $\boldsymbol{\lambda}$ is restricted to the root node of the branch-and-bound tree and they are kept unchanged for the rest of the solution process.

In order to integrate Lagrangian relaxation in the *SD-BB* Algorithm 6.1, some steps need to be tailored to the new relaxation. Concentrating on the changes resulting from the integration, the algorithm is adjusted in the following way:

Initialization

In the initialization phase, we consider the Lagrangian relaxation $d(\boldsymbol{\lambda})$ defined in (6.11), instead of $\bar{P}$. With the goal to determine good values for $\boldsymbol{\lambda}$, we apply a subgradient method for a fixed number of iterations yielding a tighter lower bound of the optimal objective function value of P. The implementation of the subgradient method is described in Section 7.3.1. Subsequently, the best computed values for the Lagrangian multipliers are fixed for the branch-and-bound process and we denote them by $\bar{\boldsymbol{\lambda}}$. Additionally, we set $d(\bar{\boldsymbol{\lambda}})$ as first lower bound LB. As mentioned above, the functions $d_k(\boldsymbol{\lambda})$ can be computed separately for constant values $\bar{\boldsymbol{\lambda}}$, which we call in the following

$$d_k(\bar{\boldsymbol{\lambda}}) = \min_{\boldsymbol{x}_k \in \mathcal{X}_k} z_k(x_k) + \sum_{n \in \mathcal{N}^k_{split}} \sum_{i \in \mathcal{I}_n} \bar{\lambda}_{ni}\, x_{ni} - \sum_{i \in \mathcal{I}_{r_k}} \bar{\lambda}_{r_k,i}\, x_{r_k,i} \qquad (L_k)$$

based on the problem formulation (6.12).

In analogy to the original relaxation $\bar{P}$ introduced in (6.6), we define the problem

$$d(\bar{\boldsymbol{\lambda}}) = \min_{\boldsymbol{x} \in \mathcal{X}} \sum_{k \in \mathcal{K}} z_k(\boldsymbol{x}_k) + \sum_{n \in \mathcal{N}_{split}} \sum_{i \in \mathcal{I}_n} \bar{\lambda}_{ni}(x_{ni} - x_{\tilde{n}i}), \qquad (\bar{L})$$

Problem Selection and Relaxation

Based on a chosen problem $P \in \mathcal{L}$ in a branch-and-bound node, we consider the corresponding Lagrangian relaxation $\bar{L}$ with subproblems L_k instead of Q_k. Since the separability of the subproblems is maintained, at most one subproblem L_l with $l \in \mathcal{K}$ needs to be solved for the determination of a new lower bound, as in case of the original relaxation.

Pruning

Considering a problem P with lower bound $LB_{\bar{L}}$ in a branch-and-bound node, the pruning by infeasibility as well as the pruning based on $LB_{\bar{L}} \geq UB$ is not concerned by the Lagrangian relaxation. In contrast, if the corresponding solution $\hat{\boldsymbol{x}}_{\bar{L}}$ satisfies all coupling constraints with respect to δ, we are not allowed to prune necessarily, as the Lagrangian relaxation modifies the objective function of the subproblems. Let $f(\boldsymbol{x})$ denote the objective function of the original problem and $g(\boldsymbol{x})$ the function of the Lagrangian relaxation. As a solution $\hat{\boldsymbol{x}}_{\bar{L}}$ is considered to be feasible if the values of the variables x_{ni} and $x_{\tilde{n}i}$ differ at most by δ, the functions $g(\hat{\boldsymbol{x}}_{\bar{L}})$ and $f(\hat{\boldsymbol{x}}_{\bar{L}})$ need not to take the same value in $\hat{\boldsymbol{x}}_{\bar{L}}$. Remember that the difference of the doubled variables is weighted with the corresponding Lagrangian multiplier in the objective function $g(\boldsymbol{x})$. Hence, in case of a feasible solution $\hat{\boldsymbol{x}}_{\bar{L}}$, the global upper bound UB is only updated if the objective function value of the original objective function $f(\boldsymbol{x})$ is less than the current upper bound UB. Additionally, the node is also pruned if a certain threshold between $f(\boldsymbol{x})$ and $g(\boldsymbol{x})$ is satisfied, i.e., we require

$$\frac{|f(\boldsymbol{x}_{\bar{L}}) - g(\boldsymbol{x}_{\bar{L}})|}{|g(\boldsymbol{x}_{\bar{L}})|} < \varepsilon,$$

for an $\varepsilon > 0$. Here, the tolerance ε is the same value which is set for the gap between the objective function value of the best feasible solution and the lower bounds of all open subproblems appearing in the termination step. Note that in the original Algorithm 6.1 the objective function of P and its relaxation $\bar{P}$ coincide making the former consideration unnecessary in that case.

Altogether, performing a few changes in the algorithm, the Lagrangian relaxation is included successfully into the branch-and-bound framework. The clear improvement in the performance of the algorithm caused by this extension is shown in Section 8.4.2.

Chapter 7

Algorithmic Implementation

This chapter is devoted to the algorithmic implementation of the *SD-BB* algorithm presented in Chapter 6. In order to make the algorithm successful, several ideas are developed which exploit either the specific properties of the algorithm or the characteristics of the *S-OPGen* problem.

The starting point of the *SD-BB* algorithm is its initialization whose core comprises the splitting of the scenario tree into several subtrees. In Section 7.1, we develop a polynomial time algorithm for a fast decomposition of the scenario tree with the objective of equally sized subtrees. Additionally, we discuss an extension of the procedure which aims at enlarging the distance of the corresponding split nodes which is favorable for the performance of the *SD-BB* algorithm. In Section 7.2, we proceed with the presentation of suitable branching techniques, where we adapt existing variable selection rules developed for LP-based branch-and-bound methods to our *SD-BB* framework. Furthermore, we describe an interval-based determination of branching points in case of continuous variables. This approach provides the basis for an efficient caching of solved subproblem during the solution process. By retaining the solutions, we avoid a redundant solving of subproblems within the computation of a lower bound, which is explained in Section 7.3. We also present a standard subgradient method for the determination of good Lagrangian multipliers in the root node of the branch-and-bound tree. Finally, Section 7.4 is dedicated to the determination of feasible solutions. We distinguish between the computation of a first solution for the *S-OPGen* problem at the beginning of the solution process and the generation of a feasible solution based on local information in a branch-and-bound node.

7.1 Decomposing a Scenario Tree

Before starting the branch-and-bound procedure, the problem needs to be reformulated based on the splitting of the scenario tree $\Gamma = (\mathcal{N}, \mathcal{E})$ into K subtrees $\Gamma_k = (\mathcal{C}_k, \mathcal{E}_k)$ with $k = \{1, \ldots, K\}$ for some $K \in \mathbb{N}$ as described in Section 6.2. Recall that by choosing a set of split nodes $\mathcal{N}_{split} \subseteq \mathcal{N}$, the subtrees are built by doubling the corresponding nodes and setting the duplicated ones as new root nodes of the resulting subtrees. Then, for each subtree Γ_k, a decomposed subproblem is formulated. Since the resulting decomposition of the original problem strongly influences the performance of the *SD-BB* algorithm, an elaborate subdivision of the scenario tree needs to be made.

More precisely, there are two major properties of the subdivision which affect the performance of the algorithm. The first one is the size of the resulting subproblems and the second one is the distance between the split nodes in the tree. Remember that a large distance of the split nodes favors a stronger independence of splitting variables corresponding to different split nodes. However, the first aspect constitutes the major impact on the performance as a strong imbalance of problem sizes most likely yields computational expensive subproblems.

Hence, in Section 7.1.1, we focus on the development of a polynomial time algorithm for the splitting of the scenario tree aiming at preferably equally sized subtrees. Having determined such a subdivision, we investigate the possibility of rearranging the split nodes with the objective of enlarging the distance among each other subject to the restriction that the subproblems are still balanced. To this end, we propose a fast heuristic method in Section 7.1.2 which performs local shifting steps of the split nodes in order to increase their distance.

7.1.1 Finding an Optimal K-Subdivision

Focusing on the size of the subtrees and thus on the size of the subproblems, we face the two tasks of choosing the number K of subtrees and selecting the split nodes comprised in $\mathcal{N}_{split}$ with $|\mathcal{N}_{split}| = K - 1$. For the decision on how to choose K and $\mathcal{N}_{split}$, the following aspects have to be taken into account: On the one hand, a large number K of subtrees leads to small subproblems which can be solved quickly. Remember that in most of the branch-and-bound nodes created during the execution of the *SD-BB* algorithm, a single subproblem needs to be solved in order to compute each lower bound. On the other hand, a large value of K results in a large number of split nodes with corresponding duplicated variables and coupling constraints defined in (6.4), which need to be restored by branching in order to obtain a feasible solution. An important aspect for the determination of a suitable K is the specific character of the problem at hand. Hence K is chosen based on a series of test runs shown in Section 8.4.1 regarding selected problem instances which are presented in Section 8.4.1.

Having chosen a fixed number K of subtrees, we follow the intuitive approach of creating subtrees whose maximum number of nodes is minimal, aiming for subproblems whose maximum size is as small as possible. In other words, we face the problem of subdividing a rooted tree Γ into K rooted subtrees $\Gamma_k = (\mathcal{C}_k, \mathcal{E}_k)$ with $k \in \{1, \ldots, K\}$ such that the maximum cardinality of the node sets $\mathcal{C}_k$ is minimized. For the subdivision, the main requirement is that each root node r_k of a subtree $\Gamma(\mathcal{C}_k)$ is also element of a further subset $\mathcal{C}_l$ with $k \neq l$, in particular r_k is a leaf in subtree $\Gamma(\mathcal{C}_l)$, except for the first root node r_1. Note that we use this reformulation in order to facilitate the notation, making the doubling of split nodes unnecessary.

In [BSP82], a related problem is studied, where a tree with weighted nodes is split into a prespecified number of subtrees requiring that the resulting node sets form a partition of $\mathcal{N}$. With the objective of minimizing the heaviest subtree, the authors provide a polynomial time algorithm for its solution. In the article [SLPS90], the objective function is changed to the minimization the imbalance between the weights of the resulting subtrees. To be more precise, the sum of the deviations between the weights of the subtrees and the average subtree weight should be as small as possible. Considering this objective function, the authors proved the problem to be $\mathcal{NP}$-complete. In contrast to the former problems, in our case the subdivision of the node set $\mathcal{N}$ is not a partition, and the subtrees have a specified form in order to be used in the *SD-BB* algorithm. Nevertheless, the problem can be solved by a polynomial time algorithm which is developed in this section. We start by formally describing the problem, using a notation which is based on the article [SLPS90].

Definition 7.1 *Let $\Gamma = (\mathcal{N}, \mathcal{E})$ be a rooted tree and let $N = |\mathcal{N}|$. By $\phi = (\mathcal{C}_1, \ldots, \mathcal{C}_K)$, we denote a K-subdivision of the node set $\mathcal{N}$ into K subsets where $K \in \{1, \ldots, N\}$ and $\mathcal{C}_k \subseteq \mathcal{N}$ for all $k \in \mathcal{K} := \{1, \ldots, K\}$. We call ϕ a* feasible K-subdivision *if the following three conditions are satisfied:*

1. *Each subgraph $\Gamma(\mathcal{C}_k)$ induced by subset $\mathcal{C}_k$ with $k \in \mathcal{K}$ is connected, i.e., $\Gamma(\mathcal{C}_k)$ is a tree.*

2. *The node sets* $\mathcal{C}_1, \mathcal{C}_2 \backslash \{r_2\}, \ldots, \mathcal{C}_K \backslash \{r_K\}$ *form a partition of* $\mathcal{N}$*, where* r_k *denotes the root node of subtree* $\Gamma(\mathcal{C}_k)$ *for all* $k \in \mathcal{K}$.

3. *For each root node* r_k *of subtree* $\Gamma(\mathcal{C}_k)$ *with* $k \in \mathcal{K}$*, all successors* $\mathcal{S}(r_k)$ *of* r_k *are comprised in* $\mathcal{C}_k$.

Without loss of generality, we assume that root node r_1 of subtree $\Gamma(\mathcal{C}_1)$ coincides with the root node of the entire tree Γ. Consequently, root r_1 takes on a special position, as it is contained in only one subset, namely $\mathcal{C}_1$. Hence, it is not removed from $\mathcal{C}_1$ for the partition considered in condition 2 of Definition 7.1. Furthermore, we define the maximum cardinality $c(\phi)$ of a subdivision $\phi = (\mathcal{C}_1, \ldots, \mathcal{C}_K)$ as

$$c(\phi) = \max_{k \in \mathcal{K}} |\mathcal{C}_k|. \tag{7.1}$$

Denoting the set of all feasible K-subdivisions by $\Phi(\Gamma, K)$, we are able to formulate the *minimum K-subdividing problem* by

$$C(\Gamma, K) = \min_{\phi \in \Phi(\Gamma, K)} c(\phi). \tag{7.2}$$

An optimal solution of Problem (7.2) is called an *optimal K-subdivision.* We remark that a feasible solution can also be described by its set of root nodes $\mathcal{R}$, comprising all root nodes r_k of subtrees Γ_k, formally $\mathcal{R} = \bigcup_{k \in \mathcal{K}} \{r_k\}$.

In order to prove that the problem can be solved by a polynomial time algorithm, we consider the related problem of finding the minimum number K^* of subsets such there exists a feasible K^*-subdivision where the cardinality of the subsets is bounded by a prespecified number U. Note that if we know how to determine a minimum K^* for a given bound U, then we can find the minimum bound U^* for a prespecified number of subsets K. This can be achieved by performing a binary search over all possible values for U yielding an optimal solution of the K-subdividing Problem (7.2). A detailed description of the algorithm is presented on page 102. In order to formulate the problem explicitly, we define the set of feasible K-subdivisions bounded by U as

$$\Pi_K(\Gamma, U) = \{\phi \in \Phi(\Gamma, K) \mid |\mathcal{C}_k| \leq U \text{ for all } k \in \mathcal{K}\}$$

with $K \in \{1, \ldots, N\}$ and $U \in \{U_{min}, \ldots, N\}$. By U_{min}, we denote the minimum value for bound U which yields a feasible problem, i.e., we define $U_{min} = \max_{n \in \mathcal{N}} \mathcal{S}(n)$, where $\mathcal{S}(n)$ comprises all successors of node $n \in \mathcal{N}$. Recall that due to condition 3 of Definition 7.1, all successors of a root node

r_i are contained in subset $\mathcal{C}_i$, which implies that at least one subset has cardinality greater or equal to U_{min}.

Now, the *tree subdividing problem bounded by U* which searches for a feasible subdivision $\phi \in \Pi_K(\Gamma, U)$ of minimum cardinality K can be described by

$$K(\Gamma, U) = \min\{K \mid K \in \{1, \ldots, N\} \text{ and } \phi \in \Pi_K(\Gamma, U)\}. \qquad (7.3)$$

If ϕ is an optimal solution of the latter problem, we call it an *optimal subdivision bounded by U*.

In the following two sections, we describe two polynomial time algorithms for the solution of Problem (7.2) and Problem (7.3). Since for the determination of an optimal K-subdivision we need to solve the problem of finding an optimal subdivision bounded by U, the next section starts with the algorithm for Problem (7.3).

Solving the Tree Subdividing Problem Bounded by *U*

For the solution of Problem (7.3) we develop a polynomial time algorithm, whose complexity is given by $\mathcal{O}(N)$. Within the algorithm we use the following notation. Instead of determining the subsets $\mathcal{C}_1, \ldots, \mathcal{C}_K$ explicitly, the algorithm searches for the corresponding root nodes $r_1, \ldots, r_K$ which are comprised in the set $\mathcal{R}$. By ω_n we refer to the cardinality of the subtree rooted in node n which is built within the tree Γ. Note that during the execution of the algorithm, Γ is modified by deletion of nodes and hence ω_n varies and needs to be updated. Following the notation of the previous chapters, $t(n)$ denotes the level $i \in \{1, \ldots, T\}$ of node $n \in \mathcal{N}$, where the level of the root node equals one and the maximum level is denoted by T. The set $\mathcal{N}_i$ contains all nodes corresponding to level i.

The underlying idea of the algorithm as well as the proof of validity are derived from the article [KM77], where the related problem of partitioning a tree into a minimum number of subtrees is studied. Each node having a positive weight, the authors require that the resulting weights of the subtrees are less than or equal to a prespecified bound. In contrast to our problem, the subtrees are induced by a partition of the node set $\mathcal{N}$, allowing the construction of subtrees upon removal of one edge of the tree. The adaptation of the algorithm for the solution of Problem (7.3) is formally described in Algorithm 7.2.

Basically, the method traverses iteratively through all nodes $n \in \mathcal{N}$ searching for root nodes creating an optimal subdivision bounded by U. In line 1, it starts by initializing all leaf nodes n with $\omega_n = 1$, indicating that a subtree

Algorithm 7.2 *Algorithm for an optimal subdivision bounded by U*

Input: Rooted tree $\Gamma = (\mathcal{N}, \mathcal{E})$ and a bound $U \in \{U_{min}, \dots, N\}$
Output: Set of roots $\mathcal{R}$ defining an optimal subdivision bounded by U

1 Set $\omega_n = 1$ for all leaf nodes in Γ
2 **for** $i = T$ down to 1 **do**
3   Set $\mathcal{N}_i = \{n \in \mathcal{N} \mid t(n) = i\}$
4   **while** $\mathcal{N}_i \neq \emptyset$ **do**
5   Select a node $n \in \mathcal{N}_i$
6   **while** $\omega_n = \sum_{v \in \mathcal{S}(n)} \omega_v + 1 > U$ **do**
7   Select a successor $v \in S(n)$ with maximum ω_v
8   Add node v to $\mathcal{R}$
9   Remove all successors $q \in \mathcal{S}(v)$ from $\mathcal{N}$ and update $\omega_v = 1$
10   **end**
11   Set $\mathcal{N}_i = \mathcal{N}_i \backslash \{n\}$.
12   **end**
13 **end**
14 Add r_1 to $\mathcal{R}$ and
15 **return** set of root nodes $\mathcal{R}$

rooted in a leaf consists of exactly one node. Then, the algorithm iterates over all time stages $i \in \{1, \dots, T\}$, starting with the maximum level T, see line 2. Having selected a node n of the current stage i, in line 6 it is verified whether the node set of the resulting subtree rooted in n exceeds the upper bound U. If this is true, a successor $v \in \mathcal{S}(n)$ with maximum ω_n is selected and added to the root node set $\mathcal{R}$. Additionally, all successors $s \in \mathcal{S}(v)$ of v are removed from the tree and ω_v is set to one. Node v is not deleted itself as it is required that any root node except for r_1 is contained in two subsets, see condition 2 of Definition 7.1. If ω_n still exceeds the upper bound U, the selection of the heaviest successor with corresponding deletion is repeated until $\omega_n \leq U$. Note that this condition can always be satisfied as $U \geq U_{min}$ is chosen. Then, a new node corresponding to stage i is selected until all nodes of i are processed. As soon as the nodes of all stages are traversed, i.e., $i = 1$, the root node r_1 is added to $\mathcal{R}$. Recall that r_1 denotes the root node of the original tree Γ and consequently is always the root node of the first subtree $\Gamma(\mathcal{C}_1)$. Finally, the set of root nodes $\mathcal{R}$ is returned which defines an optimal subdivision bounded by U.

Altogether, in Algorithm 7.2 the processing of one node $n \in \mathcal{N}$ requires a search for the heaviest successors until a given limit is reached which is achieved in at most $\mathcal{O}(|\mathcal{S}(n)|)$ steps, see [KM77]. As each node is processed once during the solution process, the algorithm has an overall running time of $\mathcal{O}(N)$.

In order to prove the optimality of the solution found by Algorithm 7.2, we use the two properties of an optimal subdivision bounded by U which are specified in Lemma 7.3 and 7.4. We define the set $\mathcal{N}(q)$ for a node $q \in \mathcal{N}$ which contains all nodes of the subtree rooted in q and the set $\overline{\mathcal{N}}(q) = (\mathcal{N} \setminus \mathcal{N}(q)) \cup \{q\}$ comprising all remaining nodes united with q. For short, the induced subtrees $\Gamma(\mathcal{N}(q))$ and $\Gamma(\overline{\mathcal{N}}_q)$ are denoted by Γ_q and $\overline{\Gamma}_q$. Now, we can formulate the first lemma showing that if a node q of the optimal root node set is known, the resulting subtrees Γ_q and $\overline{\Gamma}_q$ can be subdivided independently yielding also an optimal subdivision when combined.

Lemma 7.3 *Let $q \in \mathcal{N}$ be a root node of an optimal subdivision bounded by U of tree Γ. If $\mathcal{R}_q$ and $\overline{\mathcal{R}}_q$ are sets of root nodes defining optimal subdivisions bounded by U for the trees Γ_q and $\overline{\Gamma}_q$, respectively, then $\mathcal{R} = \mathcal{R}_q \cup \overline{\mathcal{R}}_q$ defines an optimal subdivision bounded by U for Γ.*

Proof. As root node q is element of $\mathcal{R}_q$, the union of $\mathcal{R}_q$ and $\overline{\mathcal{R}}_q$ defines subtrees of Γ whose node sets are still bounded by U and hence, $\mathcal{R}$ defines a feasible subdivision bounded by U for Γ.
In order to prove optimality, let root node set $\mathcal{R}^*$ define an optimal subdivision bounded by U for tree Γ with $q \in \mathcal{R}^*$. By $\mathcal{R}_q^*$ we denote the subset of root nodes corresponding to subtree Γ_q, i.e., $\mathcal{R}_q^* = \mathcal{R}^* \cap \mathcal{N}(q)$ and by $\overline{\mathcal{R}}_q^*$ we refer to all remaining roots, i.e., $\overline{\mathcal{R}}_q^* = \mathcal{R}^* \setminus \mathcal{R}_q^*$. Knowing that $|\mathcal{R}_q|$ is minimal for subtree Γ_q, we obtain $|\mathcal{R}_q| \leq |\mathcal{R}_q^*|$. As $\overline{\mathcal{R}}_q^*$ defines a feasible subdivision bounded by U for $\overline{\Gamma}_q$, it also follows that $|\overline{\mathcal{R}}_q| \leq |\overline{\mathcal{R}}_q^*|$. Altogether, we obtain $|\mathcal{R}| \leq |\mathcal{R}_q| + |\overline{\mathcal{R}}_q| \leq |\mathcal{R}_q^*| + |\overline{\mathcal{R}}_q^*| = |\mathcal{R}^*|$, proving optimality. □

The next lemma specifies under which conditions a node $q \in \mathcal{N}$ is element of a root node set defining an optimal subdivision bounded by U of a tree Γ.

Lemma 7.4 *Let q be a node of tree Γ and let bound $U \in \{U_{min}, \dots, N\}$. If $\omega_q > U$ and $\omega_s \leq U$ for all its successors $s \in \mathcal{S}(q)$, then there exists an optimal solution of Problem (7.3) for which the heaviest son $s_{max} = \text{argmax}_{s \in \mathcal{S}(q)} \omega_q$ of node q is contained in the corresponding root node set $\mathcal{R}$.*

Proof. Let q be a node in $\mathcal{N}$ such that the cardinality of node set $\mathcal{N}(q)$ exceeds the prespecified bound U, i.e., $\omega_q > U$. Then, we know that in an optimal subdivision of tree Γ at least one node $\bar{v} \in \mathcal{N}(q) \setminus \{q\}$ is contained in the corresponding root set $\mathcal{R}^*$, i.e., $\bar{v} \in \mathcal{R}^*$. Consequently, there exists a successor $\bar{s} \in \mathcal{S}(q)$ such that $\bar{v} \in \mathcal{N}(s)$. If $\bar{v} \neq \bar{s}$, the set $\mathcal{R} = (\mathcal{R}^* \setminus \{\bar{v}\}) \cup \{\bar{s}\}$ still defines an optimal subdivision of Γ. Since ω_s is lower than or equal to U by assumption, the resulting subdivision is still feasible, and as $|\mathcal{R}| = |\mathcal{R}^*|$ it is also optimal. Consequently, node $\bar{v}$ can be replaced by $\bar{s}$ and in the following, we may assume that $\bar{v} \in \mathcal{S}(q)$. If $\bar{v} \neq s_{max}$, i.e., $\bar{v}$ is not the heaviest son of q, root node $\bar{v}$ is replaced by s_{max}. In particular, we set $\mathcal{R} = (\mathcal{R}^* \setminus \{v\}) \cup \{s_{max}\}$, yielding an optimal subdivision bounded by U. □

Altogether, Lemma 7.4 and Lemma 7.3 ensure that the successive selection of nodes, as proposed in line 7 of Algorithm 7.2, satisfying the condition described in line 6, yields a set of root nodes which defines an optimal subdivision bounded by U.

Solving the K-Subdividing Problem of a Tree

In this section, we describe an $\mathcal{O}(N \log N)$ algorithm which finds an optimal K-subdivision of a tree, as defined in Problem (7.2). The basic idea of the algorithm is to perform a binary search over all possible objective function values, ranging from U_{min} to N. Remember that $U_{min} = \max_{n \in \mathcal{N}} |\mathcal{S}(n)| + 1$ represents the minimum upper bound for the cardinality of the resulting subsets, as each subtree $\Gamma(\mathcal{C}_k)$ with $k \in \mathcal{K}$ has to contain at least all successors $\mathcal{S}(r_k)$ of the corresponding root node r_k. Within the binary search, we make use of Algorithm 7.2, which determines a subdivision bounded by U of minimum cardinality K, where U denotes a predefined upper bound. Formally, the method of finding an optimal K-subdivision is described in Algorithm 7.5.

In detail, Algorithm 7.5 starts with the initialization of the parameters LB and UB by the values U_{min} and N, respectively. Clearly, the optimal objective function value $C(\Gamma, K)$ defined in (7.2) must be contained in the resulting set of integers $\{LB, \ldots, UB\}$. Note that by requiring $K \in \{1, \ldots, N\}$, there always exists a feasible subdivision $\phi \in \Phi(\Gamma, K)$ which can be constructed by choosing K elements of $\mathcal{N}$ and defining them as root nodes of the resulting subtrees.

Algorithm 7.5 *Algorithm for an optimal K-subdivision*

Input: Rooted tree $\Gamma = (\mathcal{N}, \mathcal{E})$ and a number $K \in \{1, \dots, N\}$ of subsets
Output: Set of roots $\mathcal{R}$ defining an optimal K-subdivision

1 Set $LB = U_{min}, UB = N$ and $\mathcal{R} = \emptyset$
2 **while** $LB < UB$ **do**
3   Set $U = \lfloor \frac{LB+UB}{2} \rfloor$
4   Solve Problem (7.3) applying Algorithm 7.2 and
5   let $\mathcal{R}^*$ be its optimal root node set of cardinality K^*
6   **if** $K^* \leq K$ **then**
7   Set $UB = U$ and $\mathcal{R} = \mathcal{R}^*$
8   **else**
9   Set $LB = U + 1$
10   **end**
11 **end**
12 **while** $|\mathcal{R}| < K$ **do**
13   Select a node $n \in \mathcal{N} \setminus \mathcal{R}$ and add it to $\mathcal{R}$
14 **end**
15 **return** root nodes $\mathcal{R}$

In the main while loop, starting in line 2, a binary search procedure over all values of the integer set $\{LB, \dots, UB\}$ is performed. The main task performed during one iteration is to verify whether the optimal function value is contained in $\{LB, \dots, U\}$ or in $\{U + 1, \dots, UB\}$ where U is the rounded mean value of LB and UB, computed in line 3. By halving the set in each iteration, i.e., updating UB or LB, the while loop is terminated if LB and UB coincide, i.e., the set $\{LB, \dots, UB\}$ includes only one element. In detail, Problem (7.3) is solved for the computed upper bound U yielding a minimum number K^* of subsets with corresponding root node set $\mathcal{R}^*$, see line 4. If K^* is less than or equal to the desired number of subsets K, a feasible K-subdivision bounded by U is found. Note that in case $K^* < K$, a feasible K-subdivision can be constructed by selecting $K - K^*$ nodes of set $\mathcal{N} \setminus \mathcal{R}^*$. All resulting subsets $\mathcal{C}_1, \dots, \mathcal{C}_K$ still satisfy $|\mathcal{C}_k| \leq U$ for all $k \in \mathcal{K}$. Hence, the root set $\mathcal{R}^*$ is stored in $\mathcal{R}$. In order to check whether there is a smaller bound U for which a feasible K-subdivision bounded by U exists, we set $UB = U$ restricting the search to $\{LB, \dots, U\}$. If $K^* > K$, then there is no feasible K-subdivision bounded by U, i.e., U is too small. Consequently, the optimal objective function value $C(\Gamma, K)$ is strictly greater than U and hence, we set $LB = U + 1$.

Having finished the while loop, in line 12 the best set of roots $\mathcal{R}$ is expanded until $|\mathcal{R}| = K$ yielding a feasible K-subdivision bounded by the minimum upper bound U. Finally, the optimal set of roots is returned.

As mentioned in Section 7.1.1, the solution of Problem (7.3) requires at most $\mathcal{O}(N)$ steps, where N denotes the number of nodes of tree Γ. Within the binary search this problem needs to be solved at most $\log(N)$ times. Thus, the complexity of Algorithm 7.5 is $\mathcal{O}(N \log N)$.

7.1.2 Rearranging an Optimal K-Subdivision

Besides the minimization of the maximal subtree addressed in the previous section, we are also interested in arranging the root nodes in such a way that the distance between each other is not too small. So far, all optimal solutions appear equally good, but for a good performance of the *SD-BB* algorithm, those root sets with larger distances are preferred. Remember that in order to exploit the basic idea of the *SD-BB* algorithm, it is desirable to choose split nodes whose influence among each other is negligible. Thus, the feasibility of the decomposed problem can be restored locally.

Transferred to the K-subdividing problem (7.2), we follow the approach of searching for optimal solutions such that the distance between pairs of root nodes is greater than a certain threshold d_{min}. To be more precise, given a predefined number K of subtrees and the optimal value U of the K-subdividing problem, we want to find a subdivision $(\mathcal{C}_1, \ldots, \mathcal{C}_K)$ satisfying the following conditions:

1. Subsets $\mathcal{C}_1, \ldots, \mathcal{C}_K$ define a feasible K-subdivision of tree Γ.
2. Each subset $\mathcal{C}_k$ with $k \in \mathcal{K}$ contains at most U nodes, where U is the optimal value of problem (7.2), i.e., $|\mathcal{C}_k| \leq U$ for all $k \in \mathcal{K}$.
3. The distance between pairs of roots is greater than d_{min}, i.e., $l(r_i, r_j) > d_{min}$ for all pairs (r_i, r_j) with $i, j \in \mathcal{K}$.

Remember that r_k represents the root node of the subtree $\Gamma(\mathcal{C}_k)$ induced by subset $\mathcal{C}_k$ and $l(r_i, r_j)$ denotes the path length between the pair of nodes (r_i, r_j), i.e., the number of edges of the r_i-r_j-path.

Motivated by the observation that condition 3 easily leads to an infeasible problem, we consider this condition as a preference rather than a constraint which has to be satisfied necessarily. In other words, the violation of condition 3 should be avoided as far as possible but may be relaxed if required.

As the minimal distance d_{min} between roots is not a condition precedent to the execution of the *SD-BB* algorithm, this treatment allows us to integrate condition 3 without loosing the guaranty of a feasible initialization problem.

Knowing that an optimal solution of the K-subdividing problem can be found quickly, we want to exploit the possibility of shifting the root nodes found by Algorithm 7.5 in order to determine a solution which satisfies condition 3. Therefore, we propose a simple shifting procedure which aims at a rearrangement of the root nodes such that the minimum distance between the roots exceeds a predefined threshold d_{min}. The procedure systematically looks at roots having a minimum distance below d_{min} and determines whether the shifting to an adjacent node is able to increase its minimum distance preserving the satisfaction of condition 1 and 2.

For a detailed description of the procedure, we make use of the following notation. For all root nodes $r \in \mathcal{R}$, the function $d(r)$ represents the minimum distance to any other root node, formally

$$d(r) = \min_{v \in \mathcal{R} \setminus \{r\}} l(r, v).$$

On this basis, we define the set $\mathcal{W}_j$ for all $j \in \{1, \ldots, d_{min}\}$ which contains all roots $r \in \mathcal{R}$ whose minimum distance $d(r)$ equals the value j, which means

$$\mathcal{W}_j = \{r \in \mathcal{R} \,|\, d(r) = j\}.$$

Additionally, let set $\mathcal{D}_r$ comprise all adjacent nodes of root r to which r can be potentially shifted, i.e.,

$$\mathcal{D}_r = \{n \in \mathcal{N} \setminus \mathcal{R} \,|\, n \in \mathcal{S}(r) \text{ or } n = p(r)\},$$

where $\mathcal{S}(r)$ denotes the set of all successors and $p(r)$ the predecessor of r. As mentioned above, the main procedure is characterized by the basic step of moving a selected root node to the predecessor or one of the successor nodes to alter the current solution. This step is summarized in Procedure 7.6.

In line 1 of the procedure, set $\mathcal{D}$ is initialized by $\mathcal{D}_r$ consisting of all adjacent nodes of node r which are not root nodes. In the while loop starting in line 2, it is explored successively whether there is an adjacent node suitable for a shifting step. The while loop terminates if all nodes of $\mathcal{D}$ are processed or if a shifting step is performed, i.e., node r is not a root node anymore. In order to shift node r to an adjacent node $q \in \mathcal{D}$, it must be ensured that the new root set $(\mathcal{R} \setminus \{r\}) \cup \{v\}$ still defines a K-subdivision bounded by U, i.e., all resulting subsets $\mathcal{C}_k$ with $k \in \mathcal{K}$ satisfy $|\mathcal{C}_k| \leq U$. Additionally,

Procedure 7.6 *shift_root_node*(Γ,$\mathcal{R}$,U,r)

Input: Rooted tree $\Gamma = (\mathcal{N}, \mathcal{E})$, root set $\mathcal{R}$ and node $r \in \mathcal{R}$
Output: Optimal K-subdivision with root set $\mathcal{R}$

```
Set D = D_r
while D ≠ ∅ and r is root node do
    Select a node q ∈ D
    if φ((R\{r}) ∪ {q}) is bounded by U and d(q) > d(r) then
        set R = R \ {r} ∪ {q}
    end
    Remove q from D
end
return root nodes R
```

it is required that $d(q)$ is strictly greater than $d(r)$ aiming at root nodes with an augmented minimum distance. We remark that if $d(q) > d(r)$, the shifting does not decrease the minimum distance of any root $v \in \mathcal{R}$ with $d(v) \leq d(q)$.

Using Procedure 7.6, we are able to describe the entire method for the increase of the minimum distance between the root nodes with the goal of exceeding the threshold d_{min}. In summary, the method is presented in Procedure 7.7 which is specified in the following.

Procedure 7.7 starts with a K-subdivision bounded by U with corresponding root set $\mathcal{R}$. In particular, $\mathcal{R}$ defines an optimal solution of the K-subdividing problem with objective value U. On this basis, the sets $\mathcal{W}_j$ with $j \in \{1, \ldots, d_{min}\}$ are created containing all roots with a minimum distance equal to j. Having sorted the critical root nodes with respect to their minimum distance, we start with those roots showing the smallest minimum distance. Therefore, we iterate over all $i \in \{1, \ldots, d_{min}\}$ represented by the outer for loop in line 3. The inner for loop is performed in order to be capable to reconsider roots with smaller minimum distance as currently indicated by index i. In detail, it may occur that the shifting of a root becomes possible after a root of larger minimum distance has been shifted. Thus, after node set $\mathcal{W}_j$ with $j = i$ is processed, all sets $\mathcal{W}_j$ with $j < i$ are explored again. In line 6, a root $r \in \mathcal{W}_j$ is selected and Procedure 7.6 performs a shifting of node r to an adjacent node if possible. Then, root r is removed from $\mathcal{W}_j$. As the minimum distance of further nodes in $\mathcal{W}_j$ may by increased, $\mathcal{W}_j$ is updated. Note that the shifting of node r as

Procedure 7.7 *increase_minimum_distance*$(\Gamma, \mathcal{R}, U, d_{min})$

Input: Rooted tree $\Gamma = (\mathcal{N}, \mathcal{E})$, an optimal K-subdivision with root set $\mathcal{R}$ and a minimum distance threshold d_{min}
Output: Optimal K-subdivision with updated root set $\mathcal{R}$

```
1  Compute minimum distance d(r) for all r ∈ R
2  Create sets W_j = {r ∈ R | d(r) = j} for all j ∈ {1,...,d_min}
3  for i = 1 to d_min do
4      for j = i down to 1 do
5          while W_j ≠ ∅ do
6              Select a root r ∈ W_j
7              Perform shift_root_node(Γ, R, U, r) returning root set R
8              Remove r from W_j
9              Remove all roots v from W_j with d(v) > j
10         end
11     end
12     Update sets W_k for all k ∈ {1,...,d_min}
13 end
14 return root nodes R
```

described in Procedure 7.6 does not affect the root sets $\mathcal{W}_k$ with $k < j$ and hence, they do not need to be updated during the execution of the inner for loop. However, the shifting of a root $r \in \mathcal{W}_j$ enlarges the cardinality of set $\mathcal{W}_{j+1}$ at least by one if $j < d_{min} - 1$. Additionally, the cardinality of root set $\mathcal{W}_k$ with $k > j + 1$ may also grow. Hence, all node sets $\mathcal{W}_k$ with $k \in \{1, \ldots, d_{min}\}$ are updated in line 12 after the inner for loop is terminated. Finally, the procedure returns a root node set $\mathcal{R}$ which still satisfies condition 1 and 2, i.e., it defines an optimal solution of the K-subdividing problem.

Altogether, the procedure terminates in a finite number of iterations as in each iteration of the outer for loop at most K nodes are processed. To be more precise, we know that $|\bigcup_{j=1}^{i} \mathcal{W}_j| \leq K$ for a fixed i as all $\mathcal{W}_j$ are disjoint subsets of the entire root set $\mathcal{R}$. As the subsets are processed in reverse order in the inner for loop, i.e., starting with $j = i$, the sets considered later in the loop are not enlarged during the execution, as explained above. Taking this into account, it is easy to see that the Procedure 7.7 is polynomial.

As the method performs only local shifting steps of root nodes, we can not guarantee to find a solution with the desired property $d(r) > d_{min}$ for all $r \in \mathcal{R}$ if it exists. However, computational studies have shown that for the scenario trees appearing in test instances of the *S-OPGen* problem, in most of the cases this method is able to rearrange the root sets to solutions with larger minimum distances. Next to the small maximum degree of the scenario trees, the relatively low value of the minimum distance d_{min} is favorable for a good performance of Procedure 7.7.

7.2 Branching

In this section, we discuss the choice of appropriate branching techniques for the *SD-BB* algorithm focusing on the impact on the performance of the solution process. In particular, we concentrate on the choice of the variable to branch on by proposing and discussing two basic concepts for the selection. The first rule consists of selecting a variable which produces the maximum violation of the coupling constraints whereas the second rule performs a one-step look-ahead for each variable based on the idea of strong branching. In case of branching on continuous variables, the choice of appropriate branching points becomes an important aspect which is discussed afterwards. We present a priority based procedure for the generation of the branching points, which allows the exploitation of the specific characteristics of the algorithm. For an overview on branching rules for MILP, see e.g. [AKM05] and [GL06].

7.2.1 Variable Selection

The choice of a branching variable directly influences the structure of the branch-and-bound tree and hence, is strongly correlated to the performance of the solution process. Aiming at a reasonable running time, we want to define a branching rule which yields a low number of branch-and-bound nodes to be evaluated.

As indicated in Section 6.3, the major task of the branching step is to tighten the lower bound on the optimal function value. Remember that in the *SD-BB* algorithm, the generation of two subproblems is done by branching on a pair of variables associated with a split node $n \in \mathcal{N}_{split}$ and corresponding the duplicated node $\tilde{n}$, i.e., on variables $(x_{ni}, x_{\tilde{n}i})$ for some $i \in \mathcal{I}_n$.

Maximum Violation Branching

The first branching rule proposed here is based on the violation of the coupling constraints which are relaxed for the decomposition. Indeed, in order to generate solutions of the decomposed problem which are also feasible for the original problem, the values of the original and of duplicated variable need to become almost identical. Hence, we follow the intuitive approach of branching on the pair of variables whose values cause the strongest violation of corresponding coupling constraints. Formally, the violation v_{ni} of split node $n \in \mathcal{N}_{split}$ and index $i \in \mathcal{I}_n$ is computed by

$$v_{ni} = |x_{ni} - x_{\tilde{n}i}|.$$

If the maximum violation is taken by several pairs of variables, the pair which has been examined first is chosen. This rule is only applied for the choice of continuous variables, as in case of binary variables the violation is either zero or one as all subproblems have been solved to optimality. Due to the special structure of the *S-OPGen* problem, the branching on binary variables has a strong impact on the solutions obtained in the two resulting subproblems. Recall that these binary variables correspond to switching processes of facilities. Additionally motivated by a series of test runs, we prioritize binary variables over the continuous ones in the branching process. Among all binary variables which violate the splitting constraint, a pair is chosen arbitrarily.

The extension of the *SD-BB* algorithm to the application of Lagrangian relaxation offers the possibility of additionally taking the Lagrangian multiplier λ into account. More precisely, the pair of variables is chosen for branching whose violation of the coupling constraint weighted with the corresponding Lagrangian multiplier is maximum. Formally, the score value v_{ni}^{λ} is computed by

$$v_{ni}^{\lambda} = |\lambda_{ni} (x_{ni} - x_{\tilde{n}i})|.$$

The prioritization of the binary variables is maintained for this rule. The heuristic motivation behind this approach is based on the appearance of product of the Lagrangian multiplier and the coupling constraint in the extended objective function. Hence, the selection of the pair of variables with maximum v_{ni}^{λ} may increase the lower bound more than taking only the maximum violation into account. A computational comparison of both alternatives is given in Section 8.4.4.

Although the general performance of the maximum violation approach within an LP-based branch-and-bound framework is not superior to the random selection [AKM05], we decide to apply this branching rule in the *SD-BB* algorithm as numerical results indicate a good performance for the solution of the *S-OPGen* problem and this common rule provides the advantage of quickly deciding on which variable to branch on.

Strong Branching

A second, more sophisticated approach is based on the concept of strong branching. Rather than taking the violation of the splitting constraints into account, this rule aims at maximally increasing the values of the objective function of the two created subproblems in a branch-and-bound node. Strong branching was originally developed by the authors of [ABCC07] for solving traveling salesman problems, where for each branching candidate two linear problems need to be solved. The rule performs a one step look-ahead by simulating the branching on possible candidates in order to compute how the objective function would change.

In order be able to compare the potential candidates based on the change in the objective function of the two resulting subproblems, typically for each candidate a score function is computed, see e.g. [LS99], which is given by

$$\text{score}(y, z) = (1 - \mu) \min\{y, z\} + \mu \max\{y, z\}, \tag{7.4}$$

Here, y and z usually represent the change in the objective function values for the generated subproblems and μ is a weighting factor between zero and one. The score function is motivated by the fact to rather branch on a variable that produces an increase in both subproblems than increasing the value of only one subproblem drastically without notably enhancing the other one as well.

In the following, the basic idea of strong branching is transferred to the *SD-BB* algorithm, tailoring the method to the specific characteristics and requirements of the algorithm. At first, we remark that a direct application of full strong branching most likely results in long computational times for the following reason: In the worst case, $2|\mathcal{N}_{split}| \sum_{n \in \mathcal{N}_{split}} |\mathcal{I}_n|$ subproblems have to be solved in each branch-and-bound node in order to decide on which variable to branch on. In contrast to the strong branching in an LP-based branch-and-bound framework, we need to solve subproblems which still impose binary restrictions. Hence, we decide to compute a good estimate for the decrease of the objective functions corresponding to the two subproblems, since a fast computation is obviously desirable.

Taking our computational experiences of the solution of the *S-OPGen* problem into account, we apply the following approach: In order to measure the change in the objective function, we decide to solve the linear relaxation of the affected subproblem instead of solving the original one to optimality. This approach can be compared with execution of only a few simplex iterations in the LP-based case. Motivated by the observation that the linear relaxation provides nearly optimal solutions, most likely the change in the objective function corresponding to the linear relaxation gives a good estimate.

As the computational effort in each branch-and-bound node is still high, we make use of a common speedup possibility which restricts the set of potential candidates for branching to a subset $\mathcal{C}$ of selected variables. To be more precise, we follow the approach of distinguishing between binary and continuous variables for the following reasons. Numerical results have indicated that in the case of the $S-OPGen$ problem the branching on binaries has a greater impact on the lower bound than branching on continuous ones. Additionally, the recovery of a relaxed coupling condition in case of a binary variable can be achieved within one branching step, whereas for continuous variables typically a series of branching steps are necessary. Hence, we decide to prioritize binary variables for branching and to restrict the strong branching approach to continuous variables as they pose a greater challenge during the execution. Formally, for a split node $n \in \mathcal{N}_{split}$, the set $\mathcal{C}_n$ of continuous variables is defined by

$$\mathcal{C}_n = \{i \in \mathcal{I}_n \,|\, x_{ni} \text{ is continuous}\}$$

and we set $\mathcal{C} = \bigcup_{n \in \mathcal{N}_{split}} \mathcal{C}_n$. Hence the number of LPs to be solved reduces to $2|\mathcal{C}|$.

Having chosen a reduced set of potential candidates for branching, we can now specify the branching step itself using the notation introduced in Section 6.4. Therefore, we combine the strong branching idea with the two speed up techniques explained above. Let $\bar{\boldsymbol{x}}$ denote the optimal solution of the relaxed problem obtained in a branch-and-bound node. Then, the set of all branching candidates is given by $\mathcal{D} = \bigcup_{n \in \mathcal{N}_{split}} \mathcal{D}_n$ with

$$\mathcal{D}_n = \{i \in \mathcal{I}_n \cap C_n \mid \bar{x}_{ni} - \bar{x}_{\tilde{n}i} > \delta\}.$$

This means that the set comprises the indices of all splitting variables corresponding to split node $n \in \mathcal{N}_{split}$ which are element of the restricted subset C_n and which violate a splitting constraint. Recall that δ represents the accuracy assumed for the *SD-BB* algorithm. Then, for all pairs of candidates

$(x_{ni}, x_{\tilde{n}i})$ with $n \in \mathcal{N}_{split}$ and $i \in \mathcal{D}_n$, we compute the score value v_{ni} using score function (7.4). In particular, in order to compute the degradation of the objective function caused by branching in one of the two successor nodes, only the affected subproblem L_k for an $k \in \mathcal{K}$ has to be considered. Recall that in each branch-and-bound node at most one subproblem L_k needs to be resolved for the determination of a lower bound, as explained in Section 6.3. Hence, the violation v_{ni} is computed by

$$v_{ni} = \begin{cases} \text{score}(z_{\kappa(n)} - z^+_{\kappa(n)}, z_{\kappa(\tilde{n})} - z^-_{\kappa(\tilde{n})}), & \text{if } L_{\kappa(n)} \text{ is affected in the right subproblem and } L_{\kappa(\tilde{n})} \text{ in the left one,} \\ \text{score}(z_{\kappa(\tilde{n})} - z^+_{\kappa(\tilde{n})}, z_{\kappa(n)} - z^-_{\kappa(n)}), & \text{otherwise,} \end{cases}$$

where $\kappa(n)$ represents the index of the subtree to which node n belongs. Furthermore,

$z_{\kappa(n)}$ denotes the optimal function value of the LP relaxation of subproblem $L_{\kappa(n)}$ in the current branch-and-bound node.

$z^+_{\kappa(n)}$ denotes the optimal function value of the LP relaxation of the right subproblem $L_{\kappa(n)}$ after branching on the pair $(x_{ni}, x_{\tilde{n}i})$.

$z^-_{\kappa(n)}$ denotes the optimal function value of the LP relaxation of the left subproblem $L_{\kappa(n)}$ after branching on the pair $(x_{ni}, x_{\tilde{n}i})$.

The values $z_{\kappa(\tilde{n})}, z^+_{\kappa(\tilde{n})}$, and $z^-_{\kappa(\tilde{n})}$ are defined analogously with respect to the duplicated node $\tilde{n}$.

Finally, we remark that in case the LP relaxation of a subproblem in one of the successor nodes is infeasible, we can deduce that the entire problem in the resulting branch-and-bound node is infeasible as well. Applying a relaxation instead of an approximation for the estimation of the branching impact, the resulting branch-and-bound node can be pruned immediately.

In summary, the branching rule presented in this section aims at a reduction of the overall number of branching nodes by taking the estimated change in the objective function of the two subproblems into account. By simulating the branching step on continuous variables, we hope to cut nodes earlier in the solution process yielding a better performance of the *SD-BB* algorithm. Hence, this branching rule and the maximal violation branching rule are computationally investigated and the numerical results are presented in Section 8.4.4.

7.2.2 Branching on Continuous Variables

In contrast to the branching on binary variables, the branching on continuous ones makes the choice of adequate branching points necessary. Therefore, we consider a continuous splitting variable x_{ni} with corresponding duplicated variable $x_{\tilde{n}i}$ whose values $\bar{x}_{ni}$ and $\bar{x}_{\tilde{n}i}$ violate the coupling condition. In the following, we assume that this pair of variables $(x_{ni}, x_{\tilde{n}i})$ is chosen for the next branching step. For the determination of an appropriate branching point $b \in [x_{ni}^{min}, x_{ni}^{max}]$, it is obviously desirable that the resulting branching inequalities with bound b make the solution $\bar{\boldsymbol{x}}$ of the current branch-and-bound node infeasible. Hence, an intuitive way to choose b is to compute the midpoint of both values, formally

$$b = \frac{|\bar{x}_{ni} + \bar{x}_{\tilde{n}i}|}{2}.$$

However, in the *SD-BB* algorithm we apply an enhanced approach which is better suited for the application within the algorithm. By defining a fixed set of branching points in advance being adaptively refined if necessary, this method supports the exploitation of the caching procedure which is presented in Section 7.3.2. In detail, we want to make use of the fact that identical or closely related subproblems L_k with $k \in \mathcal{K}$ may occur in different branch-and-bound nodes during the solution process. For details we refer to Section 7.3.2. Hence, it is desirable to branch on points which are previously defined rather than computing the branching points individually in each branch-and-bound node. By the definition of a set of fixed points it is more likely to generate subproblems which may be used later in the solution process. In order to increase this probability, we start with a small set of predefined branching points. If it becomes necessary, the branching points are refined during the solution process based on a certain refinement rule.

Formally, for each continuous splitting variable x_{ni} with $n \in \mathcal{N}_{split}$ and $i \in \mathcal{I}_n$, the interval $[x_{ni}^{min}, x_{ni}^{max}]$ is subdivided into L equidistant intervals for an $L \in \mathbb{N}$ by defining $L+1$ points $x_{ni}^{min} = a_1 < \ldots < a_{L+1} = x_{ni}^{max}$. These points provide the start set of branching points for the pair of variables $(x_{ni}, x_{\tilde{n}i})$ at the beginning of the solution process. When the variables x_{ni} and $x_{\tilde{n}i}$ are chosen for the next branching step, we apply the following procedure for the determination of a branching point b.

Procedure 7.8 starts with the search for a branching point a_l out of the set $\{a_1, \ldots, a_{L+1}\}$ which is nearest to the midpoint m of the two values $\bar{x}_{ni}$

Procedure 7.8 *compute_branching_point*$((x_{ni},x_{\tilde{n}i}),(\bar{x}_{ni},\bar{x}_{\tilde{n}i}),\{a_1,\dots,a_{L+1}\})$

Input: Pair of branching variables $(x_{ni}, x_{\tilde{n}i})$ with values $(\bar{x}_{ni}, \bar{x}_{\tilde{n}i})$ and set of branching points $\{a_1, \dots, a_{L+1}\}$
Output: Branching point b

1 Choose branching point a_l with $l \in \{1, \dots, L+1\}$ being nearest to $m = \frac{\bar{x}_{ni}+\bar{x}_{\tilde{n}i}}{2}$, according to (7.5)
2 **if** $a_l \in \big(\min\{\bar{x}_{ni}, \bar{x}_{\tilde{n}i}\}, \max\{\bar{x}_{ni}, \bar{x}_{\tilde{n}i}\}\big)$ **then**
3   Set $b = a_l$ and **return** b
4 **end**
5 Determine interval $I = [a_k, a_{k+1}]$ for an $k \in \{1, \dots, L\}$ such that $\bar{x}_{ni}, \bar{x}_{\tilde{n}i} \in I$
6 Compute midpoint $a_{mid} = \frac{I_{min}+I_{max}}{2}$ of interval $I = [I_{min}, I_{max}]$
7 **while** $a_{mid} \notin (\min\{\bar{x}_{ni}, \bar{x}_{\tilde{n}i}\}, \max\{\bar{x}_{ni}, \bar{x}_{\tilde{n}i})\}$ **do**
8   Update interval I according to (7.6)
9   Compute midpoint a_{mid} of I
10 **end**
11 Set $b = a_{mid}$ and **return** b

and $\bar{x}_{\tilde{n}i}$. In detail, we search for a point a_l whose index $l \in \{1, \dots, L+1\}$ satisfies

$$l = \underset{k \in \{1,\dots,L\}}{\operatorname{argmin}} \left| a_k - \frac{\bar{x}_{ni} + \bar{x}_{\tilde{n}i}}{2} \right|. \tag{7.5}$$

If a_l lies between $\bar{x}_{ni}$ and $\bar{x}_{\tilde{n}i}$, i.e., $\min\{\bar{x}_{ni}, \bar{x}_{\tilde{n}i}\} < a_l < \max\{\bar{x}_{ni}, \bar{x}_{\tilde{n}i}\}$ holds, a suitable branching point is found. Note that a branching step using the bound a_l ensures that the current solution $\bar{\boldsymbol{x}}$ becomes infeasible in both subproblems. In this case, the procedure terminates and returns the current value b providing the branching point for the next branching step. Otherwise, if there is no predefined branching point between both values, there exists an interval $I = [a_k, a_{k+1}]$ with $k \in \{1, \dots, L+1\}$ such that $\bar{x}_{ni}$ and $\bar{x}_{\tilde{n}i}$ are included in I, see line 6. For the interval I, the corresponding midpoint a_{mid} is computed. In line 8, the main while loop starts which is responsible for the recursive halving of the current interval I until a midpoint a_{mid} between the values $\bar{x}_{ni}$ and $\bar{x}_{\tilde{n}i}$ is found. To be more precise, in case a_{mid} lies not between $\bar{x}_{ni}$ and $\bar{x}_{\tilde{n}i}$, we compute the new interval I by

$$I = \begin{cases} [I_{min}, a_{mid}], & \text{if } \bar{x}_{ni}, \bar{x}_{\tilde{n}i} \le a_{mid} \\ [a_{mid}, I_{max}], & \text{otherwise,} \end{cases} \tag{7.6}$$

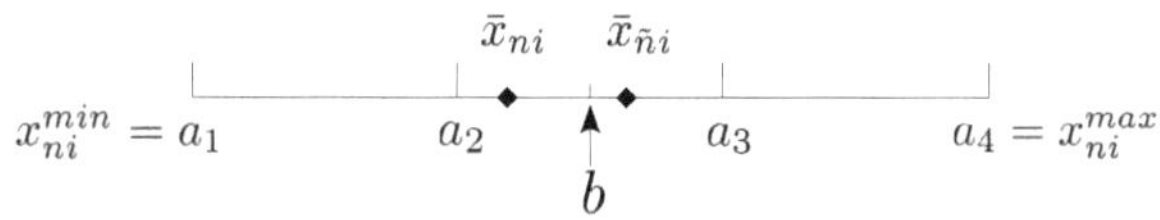

Figure 7.1: Branching on a pair of continuous variables

where I_{min} and I_{max} denote the lower and upper bound of the previous interval, respectively. As soon as a suitable branching point between $\bar{x}_{ni}$ and $\bar{x}_{\tilde{n}i}$ is found, the procedure stops and the point is returned.

The search for a branching point b is exemplarily visualized in Figure 7.1, where the initial set of branching points consists of the four points a_1, a_2, a_3 and a_4. In the picture, there is no original branching point a_k which lies between the values $\bar{x}_{ni}$ and $\bar{x}_{\tilde{n}i}$. Instead, both values are element of the interval $I = [a_2, a_3]$. Indeed, by computing the midpoint a_{mid} of I a suitable branching point $b = a_{mid}$ is obtained, since it satisfies $\bar{x}_{ni} < a_{mid} < \bar{x}_{\tilde{n}i}$. The selection of a branching point close to the midpoint m aims at balancing the violations resulting from the branching step. Although there is no guarantee for a good performance of this selection rule, it most likely yields a balanced change of the corresponding variable values in each of the resulting subproblems.

Altogether, the rule provides a reasonable and fast way of determining a suitable branching point b. By using a predefined set of branching points together with a fixed refinement strategy instead of computing branching points individually in each branch-and-bound node, this rule additionally supports the caching approach of reusing solutions obtained during the solution process, which is described in detail in Section 7.3.2.

7.3 Computing Lower Bounds

One of the fundamental parts of the *SD-BB* algorithm concerns the computation of lower bounds on the optimal objective function value. Recall that in contrast to the widespread approach of using the LP-relaxation, the lower bound in the *SD-BB* algorithm is computed by relaxing the coupling constraints. In particular, we make use of Lagrangian relaxation whose application to the *S-OPGen* problem is presented in Section 6.4. In the following two sections, we present our approaches in order to improve the computation of the lower bound aiming at a good performance of the algorithm. In detail, Section 7.3.1 focuses on the determination of a first lower bound in the root node of the branch-and-bound tree, applying a subgradient method to the Lagrangian dual function. In Section 7.3.2, we discuss a possibility of accelerating the computation of the lower bound in a branch-and-bound node in general, by making use of information obtained earlier in the solution process. To this end, a caching procedure is presented allowing the systematic storage and recovery of solutions.

7.3.1 Generation of a First Lower Bound

As explained in Section 6.2, at the beginning of the solution process a first lower bound on the optimal objective function value is computed based on the decomposition of the entire problem into subproblems. Recall that due to the relaxation of the coupling constraints, all resulting subproblems can be solved separately. By summing up the optimal objective function values of the subproblems, a first lower bound is obtained. In Section 6.4, an extension of the *SD-BB* algorithm by the application of a Lagrangian relaxation is proposed with the aim of improving the performance of the algorithm. In order to compute a tight lower bound, a suitable choice of the Lagrangian multipliers is essential. With the aim of determining good values for the multipliers, we implement a subgradient method as described e.g. in [Geo74] and [Wol98], which is specified in the following.

Based on the notation introduced in Section 6.4, let $\boldsymbol{\lambda}$ be the Lagrangian multiplier vector associated with the set of relaxed coupling constraints and let $d(\boldsymbol{\lambda})$ denote the resulting dual function as defined in (6.11). Then, a subgradient $\boldsymbol{\xi} = (\boldsymbol{\xi}_n)_{n \in \mathcal{N}_{split}}$ for the dual function d in a given point $\bar{\boldsymbol{\lambda}}$ is computed by

$$\boldsymbol{\xi}_n = (\bar{x}_{n,1} - \bar{x}_{\tilde{n},1}, \ldots, \bar{x}_{n,|\mathcal{I}_n|} - \bar{x}_{\tilde{n},|\mathcal{I}_n|}), \tag{7.7}$$

for all $n \in \mathcal{N}_{split}$, where $\bar{x}_{ni}$ and $\bar{x}_{\tilde{n}i}$ represent the optimal solution values of the corresponding variables which are obtained from the minimization of problem $d(\bar{\boldsymbol{\lambda}})$ defined in Section 6.4.2 for a fixed $\bar{\boldsymbol{\lambda}}$. Recall that $\mathcal{I}_n$ comprises all indices of time-connecting variables associated with node n. Below, the basic steps of the subgradient method for the Lagrangian relaxation used for the *S-OPGen* problem are outlined.

In detail, Algorithm 7.9 starts with the initialization of the Lagrangian multipliers. As initial values, we make use of the optimal dual solution values of the relaxed constraints obtained by solving the LP-relaxation of problem P. Computational investigations have shown that these values provide a good starting point for the subgradient method applied to the Lagrangian relaxation of the *S-OPGen* problem.

Algorithm 7.9 *Subgradient method for the S-OPGen Problem*

Input: Lagrangian dual $d(\boldsymbol{\lambda})$ defined in (6.11) and an iteration limit $\bar{R}$

Output: Lower bound LB for problem P defined in (6.5) and multipliers $\bar{\boldsymbol{\lambda}}$

Step 1: Initialization
Choose initial Lagrangian multipliers $\bar{\boldsymbol{\lambda}} \in \mathbb{R}^{|\mathcal{I}_1|} \times \ldots \times \mathbb{R}^{|\mathcal{I}_{N_s}|}$
Set iteration number $r = 0$ and step length $\mu = 2$.

Step 2: Evaluation of the Lagrangian Dual
Solve subproblems $d_k(\bar{\boldsymbol{\lambda}})$ for all $k \in \mathcal{K}$ with optimal solutions $\bar{\boldsymbol{x}}_{\boldsymbol{k}}$ and values $\bar{d}_k$ and set $LB = \sum_{k\in\mathcal{K}} \bar{d}_k$.
Compute subgradient $\boldsymbol{\xi}$ of d in $\bar{\boldsymbol{\lambda}}$ as in (7.7) using $(\bar{\boldsymbol{x}}_1, \ldots, \bar{\boldsymbol{x}}_K)$.

Step 3: Stopping
If r reaches iteration limit $\bar{R}$ or $\boldsymbol{\xi} = \mathbf{0}$ then stop and return LB and $\bar{\boldsymbol{\lambda}}$

Step 4: Updating
Compute step length μ as defined in (7.8) and set $\bar{\boldsymbol{\lambda}} = \bar{\boldsymbol{\lambda}} + \mu\,\boldsymbol{\xi}$.
Set $r = r + 1$ and go to Step 2.

The main part of the subgradient method comprises the evaluation of the dual function in each iteration r as described in Step 2. Due to the relaxation of the coupling constraints, $d(\bar{\boldsymbol{\lambda}})$ can be decomposed into K independent subproblems $d_k(\bar{\boldsymbol{\lambda}})$ defined in Section 6.4.2 yielding the optimal function values $\bar{d}_k$. Thus, the sum of all $\bar{d}_k$ with $k \in \mathcal{K}$ provides a lower bound on the optimal value of P. In order to avoid long running times and to provide a first lower bound quickly, the method is restricted to a fixed number of iterations $\bar{R}$. The algorithm also terminates in case the subgradient $\boldsymbol{\xi}$ equals zero because the solution found is feasible for problem P as well as optimal.

The choice of the step length μ has a major impact on the performance of the algorithm. As we are interested in generating a good lower bound within the iteration limit $\bar{R}$ rather than focusing on the convergence of the method, we follow an approach proposed by [Fis81], which showed a good performance empirically, although is does not guaranty the convergence to the optimum. Within this step length rule, the step length μ is halved in iteration r if the method has failed to improve the function d for a predetermined number N of iterations. Formally, we set

$$\mu_r = \begin{cases} \frac{\mu_{r-1}}{2}, & \text{if } d \text{ has failed to increase for the last } N \text{ time steps,} \\ \mu_{r-1}, & \text{otherwise.} \end{cases} \tag{7.8}$$

Having chosen a suitable step length, the Lagrangian multiplier vector is updated by using the subgradient as step direction as described in Step 4.

Altogether, the subgradient method provides the possibility of tightening the lower bound by determining suitable values for the Lagrangian multipliers λ. Within the algorithm, the choice of the maximal number of iterations $\bar{R}$ has a significant impact on the quality of the solution. On the one hand, a large number of iterations clearly favors the computation of a tight lower bound which provides a good starting point for the execution of the *SD-BB* algorithm. On the other hand, many iterations may result in long running times as the evaluation of the Lagrangian dual may be expensive. Recall that in order to compute the optimal value of a problem $d(\bar{\boldsymbol{\lambda}})$, K subproblems $d_k(\bar{\boldsymbol{\lambda}})$ need to be solved. However, computational studies have shown that at the beginning of the solution process of the subgradient method the lower bound can be quickly improved whereas later in the process the improvement decreases considerably. The exact number of iterations is chosen based on a number of test runs, presented in Section 8.4.2.

7.3.2 Caching of Subproblems for the Computation of Lower Bounds

The purpose of this section is to present the caching approach for solved subproblems during the solution process of the *SD-BB* algorithm in order to avoid redundant solutions of problems. Due to the specific decomposition method, similar or identical subproblems may occur various times during the execution process which is exploited by the following caching procedure.

Besides branching, the performance of the algorithm essentially depends on the computing time for the generation of a lower bound, as it is performed in each node of the branch-and-bound tree. Thus, the basic idea is to use information obtained earlier in the solution process in order to speed up the computation. More precisely, in each branch-and-bound node one subproblem P_k is affected and we need to know the corresponding optimal solution value for the generation of the lower bound, as explained in Section 6.3. For a quick determination, we want to make use of already solved subproblems which are also suitable for P_k rather than solving the subproblem again. Clearly, only subproblems $\overline{P}_k$ with the same subtree Γ_k need to be considered. The solution of a subproblem $\overline{P}_k$ may be used for P_k if it fulfills the following conditions:

- The feasible set D_k of P_k is contained in the feasible set $\overline{D}_k$ of $\overline{P}_k$, i.e., $D_k \subseteq \overline{D}_k$.
- The optimal solution $\bar{\boldsymbol{x}}_k$ of $\overline{P}_k$ is contained in D_k, i.e., $\bar{\boldsymbol{x}}_k \in D_k$.

Satisfying these conditions, we know that $\bar{\boldsymbol{x}}_k$ is also an optimal solution of P_k, as $\overline{P}_k$ constitutes a relaxation. In particular, identical subproblems may be used. The possibility of reusing solutions later in the solution process results from the decomposition of the entire problem into independent subproblems together with the branching on pairs of variables associated with the split nodes. Exemplarily, their appearance is shown in the following instance.

Example 7.10 Consider the scenario tree Γ which is decomposed into three subtrees Γ_1, Γ_2, and Γ_3 with split nodes $\mathcal{N}_{split} = \{u, v\}$ as shown in Figure 7.2. Throughout the example, we assume that all variables are binary. On the right hand side of the figure, the corresponding branching tree is illustrated reflecting the solution process. In detail, it starts with the branching on the pair of variables $(x_u, x_{\tilde{u}})$. In the resulting branch-and-bound nodes 2 and 3, the solution values $\bar{x}_v$ and $\bar{x}_{\tilde{v}}$ violate the corresponding coupling constraint. Hence, in both branch-and-bound nodes, a branching step on the pair of variables $(x_v, x_{\tilde{v}})$ is performed yielding node 4 and 5 on the left hand side and node 6 and 7 on the right hand side. Now, if we consider the branch-and-bound nodes 4 and 6, the subproblem P_3 associated with subtree Γ_3 shows identical fixations on the corresponding splitting variables. Thus, in both cases an identical subproblem needs to be solved. By caching the solution of the subproblem which has been examined first, the second solving is rendered unnecessary which may result in a reduction of the overall running time of the solution process.

Aiming at recovering appropriate solutions as quickly as possible, we need to store and to organize the solutions in such a way that a fast retrieval is possible. Exploiting that only subproblems corresponding to the same subtree Γ_k need to be compared, the caching is performed separately for each $k \in \mathcal{K}$. On this basis, the specific data structure of the information to be stored is taken into account. Hence subsequently, we discuss the caching for a fixed $k \in \mathcal{K}$.

Comparing two subproblems P_k and $\overline{P}_k$ associated with subtree Γ_k, they can only differ in tightened bounds concerning variables x_{ni} with $i \in \mathcal{I}_n$ of a split node $n \in \Gamma_k \cap \mathcal{N}_{split}$ or in variables $x_{\tilde{n}_k i}$ with $i \in \mathcal{I}_{\tilde{n}_k}$ where $\tilde{n}_k$ is the

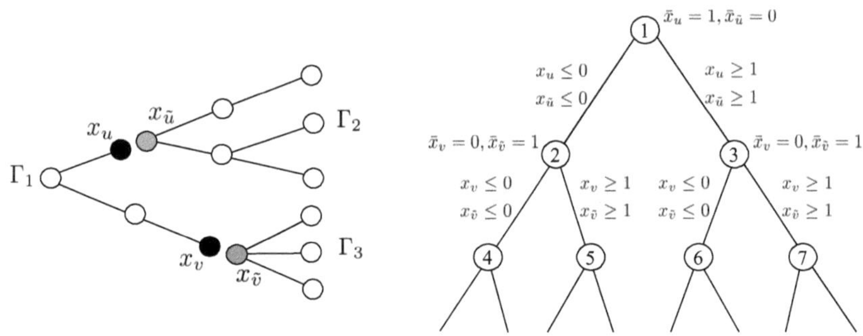

Figure 7.2: Example of identical problems occurring during the solution process

root node of Γ_k. Recall that in general, the branching is only performed on selected pairs of variables which are associated with a split node $n \in \mathcal{N}_{split}$ and the corresponding duplicated node $\tilde{n}$. For short, we call theses variables *splitting variables*. In order to compare the feasible set of the subproblems, we distinguish between binary and continuous splitting variables. The main idea is to assign a key to each subproblem P_k which is calculated based on the domain of the binary splitting variables x_{ni} resulting from branching operations. A natural encoding b for the domain of a binary variable x_{ni} is

$$b = \begin{cases} 0, & \text{if } x_{ni} \text{ is already fixed to zero,} \\ 1, & \text{if } x_{ni} \text{ is already fixed to one,} \\ 2, & \text{if } x_{ni} \text{ is not fixed, i.e., } x_{ni} \in \{0,1\}. \end{cases}$$

Hence, a key consists of a sequence of numbers $b \in \{0,1,2\}$. Let $(c_1, \ldots, c_L)$ with $L \in \mathbb{N}$ encode the domain of all binary splitting variables associated with problem P_k as described above. Then, a key $(b_1, \ldots, b_L)$ is said to be *valid* for subproblem P_k if $b_l = c_l$ or $b_l = 2$ for all $l \in \{1, \ldots, L\}$.

Each key is associated with a collection of records where each of them consists of solutions $\bar{\boldsymbol{x}}_k$ with corresponding fixations of the binary splitting variables and lower and upper bounds of the continuous splitting variables. As the fixations of the binaries define the associated key, the records comprised in one collection can only differ in the upper and lower bounds of the continuous variables and the optimal solution itself. By classifying the records by their fixations, each record is assigned to exactly one collection with corresponding key.

Thus, if we want to retrieve a stored solution $\bar{\boldsymbol{x}}_k$ which is also optimal for the current problem P_k, at first, all keys have to be traversed sequentially. If a valid key is found, we search the corresponding collection for a record with valid lower and upper bounds of the continuous variables and with a feasible stored solution. Note that in order to verify the feasibility of solution $\bar{\boldsymbol{x}}_k$ for problem P_k, only the values of the splitting variables need to be considered. In the case that no suitable record is found, the search for valid keys in the list is continued. When all keys are processed and no suitable record is found, a new record is created which is stored in the collection corresponding to the key defined by the fixations of the binary variables. If such a key does not exist yet, a new key is computed and added to the list.

This approach is appropriate for our purposes, as in our problem the majority of splitting variables are binary, which can be encoded and compared fast. Additionally, in most of the cases, the systematic caching reduces the number of records which are checked, as usually not all keys are valid for the current problem P_k and hence, records with invalid fixations are disregarded in advance. Altogether, the caching provides the possibility of improving the performance of the *SD-BB* algorithm, as, in contrast to the searching for a suitable record, the solution of a subproblem P_k may be computationally expensive since a subproblem still imposes integer restrictions.

7.4 Computing Feasible Solutions

For the algorithmic implementation concerning the computation of a feasible solution, we propose two different methods, each of them showing its own characteristics with respect to the performance and solution quality. The first approach aims at finding good quality solutions from scratch, whereas the second one makes use of information available during the solution process in order to quickly generate feasible solutions.

7.4.1 Primal Start Solution

The first heuristic aims at providing a good feasible solution independently from the solution process. For this purpose, we apply the approximate-and-fix heuristic presented in Section 5.3 which is able to provide near optimal solutions without depending on any additional information. Based on computational investigations, we decide to use the heuristic in the root node of the *SD-BB* algorithm in order to avoid the incrementation of the overall running time. For the results obtained by the computational investigations of the approximate-and-fix heuristic, we refer to Section 8.3.2.

7.4.2 Primal Solutions Based on Local Information

Next to the approximate-and-fix heuristic, we develop a second method for the construction of feasible solutions which is used in the course of the *SD-BB* algorithm after the first branch-and-bound node has been processed. The main idea of the heuristic constitutes the exploitation of the local information available in the corresponding branch-and-bound node aiming at restoring the relaxed coupling constraints. As this method is applied various times during the solution process, a fast generation is essential for the entire algorithm.

The basic concept of this approach is to fix as much binary variables as possible to values obtained from the solution of the relaxed problem in the current branch-and-bound node. Remember that the binary variables of the solution of the relaxed problem are already integer feasible. By additionally requiring the satisfaction of the coupling constraints, a new problem is created whose solutions are also feasible for the original problem. We remark that the fixation of the binary variables considerably reduces the size of the problem, allowing a fast determination of a feasible solution on the one hand, but may also result in an infeasible problem on the other hand. Both aspects have to be carefully weighed up against each other in order to create a fast and reliable heuristic. Hence, the choice of the location and of the number of binary variables to be fixed is discussed in the following.

In detail, the specific structure of the decomposition is taken into account providing the basis for deciding which variables to fix. As the subproblems have been solved under the relaxation of the coupling constraints, mainly the fixations of variables close to split nodes may lead to an infeasible problem if the coupling constraints are restored. Hence, we decide to keep the binary variables free in the surrounding of all split and duplicated nodes. Formally, we define a distance d such that all binary variables corresponding to nodes within this distance to a split node $n \in \mathcal{N}_{split}$ or a corresponding duplicated node $\tilde{n}$ are not fixed. The set of these nodes is defined by

$$\mathcal{N}_{free} = \{m \in \mathcal{N} \mid \exists n \in \mathcal{N}_{split} \text{ with } l(n,m) \leq d \text{ or } l(\tilde{n},m) \leq d\}. \quad (7.9)$$

Recall that $l(n, m)$ represents the path length between the nodes n and m. Exemplarily, the node set $\mathcal{N}_{free}$ is visualized in Figure 7.3, assuming $d = 1$.

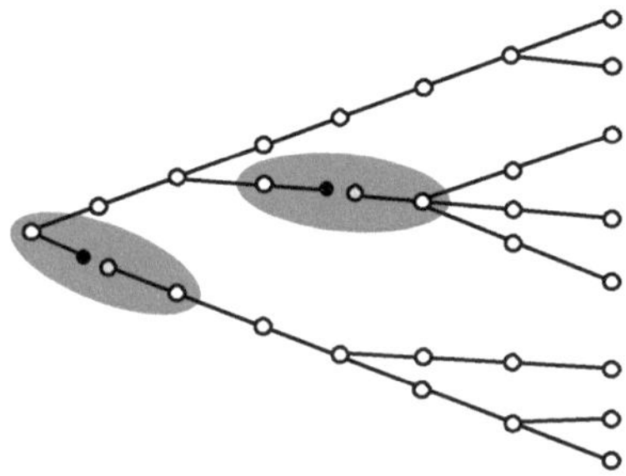

Figure 7.3: Example of a split scenario tree with fixed and free regions

In the picture, the solid nodes correspond to split nodes and the hachured nodes to the doubled ones. Then, all binary variables which correspond to nodes in the grey highlighted region are free and the binary variables of the remaining nodes are fixed to values retrieved from the solution of the decomposed problem.

For the choice of the distance parameter d, the large number of binary variables arising from a large value of d has to balanced against the possibility of creating infeasible subproblems if d is too small. The explicit choice of d is based on the specific properties of the energy system of the *S-OPGen* problem, which is discussed in Section 8.4.3.

In contrast to the specific treatment of binary variables, we decide to leave all continuous variables free in order to prevent infeasibilities. Altogether, we obtain the following reduced problem which has to be solved for the determination of a feasible solution:

$$\begin{aligned} \min \quad & \sum_{k \in \mathcal{K}} z_k(\boldsymbol{x}_k) \\ \text{s.t.} \quad & x_{ni} = x_{\tilde{n}i}, \quad \text{for } i \in \mathcal{I}_n \text{ and } n \in \mathcal{N}_{split}, \\ & x_{ni} = \bar{x}_{ni}, \quad \text{for } i \in \mathcal{B}_n \text{ and } n \in \mathcal{N} \setminus \mathcal{N}_{free}, \\ & \boldsymbol{x}^k \in \mathcal{X}_k, \qquad \text{for } k \in \mathcal{K}. \end{aligned}$$

The first set of equations represents the coupling constraints which are imposed to restore the connection of the decomposed subproblems. Set $\mathcal{B}_n$ is the index set of all binary variables of a node $n \in \mathcal{N}$ comprising the binary decision as well as the binary linearization variables. The selected node set $\mathcal{N}_{free}$ is defined in (7.9). Altogether, the second set of equations describes the fixation of the binary variables x_{ni} to the corresponding values $\bar{x}_{ni}$ which are obtained from the solution of the decomposed problem solved for the computation of the lower bound in the current branch-and-bound node. Finally, $\mathcal{X}_k$ comprises all feasible solutions of the k-th subproblem as formally defined in (6.2) of Section 6.2.

Indeed, the resulting problem is not decomposable anymore as the coupling constraints are restored, but in return it contains hardly any binary restrictions except for those corresponding to nodes of $\mathcal{N}_{free}$. As the determination of an upper bound is essential for the execution of the *SD-BB* algorithm, this procedure is executed frequently during the solution process. The specific frequency of the application is discussed in Section 8.4.3.

This concludes the description of the algorithmic implementation of the *SD-BB* algorithm providing the basis for a good performance. All presented approaches and techniques are included in the framework of the *SD-BB* algorithm and are computationally investigated and compared in the following chapter.

Chapter 8

Computational Results

In this chapter, we present a series of computational results for the solution of the *D-OPGen* and *S-OPGen* problem with the aim of documenting the performance of the solution approaches presented in the previous chapters. In Section 8.1, we start with a description of the problem instances considered for the computations by specifying the power generation system, followed by the presentation of the generated scenario trees. Subsequently, we computationally investigate the incorporation of facets determined for the stochastic switching polytope of Chapter 4 as cutting planes to a branch-and-cut algorithm. The main focus of this chapter lies on the investigation on the computational behavior of the *SD-BB* algorithm whose development and implementation has been described in the previous two chapters. In this context, we perform a systematic calibration of the applied methods and parameters with the aim of obtaining general suggestions for the setting. On this basis the algorithm is applied to instances of larger size where we scale the basic characteristics which define a problem instance of the *S-OPGen* problem.

For our implementation of the *S-OPGen* problem presented in Section 3.2.2, we use the C++ API of ILOG Concert Technology, see [CPL]. The implementation of the *SD-BB* algorithm is based on the commercial solver ILOG CPLEX 11.1 which is used to solve mixed-integer linear programs by applying a branch-and-cut algorithm. With the C++ API, CPLEX provides the possibility of using optimization callbacks in order to guide the solution process and to include user written subroutines. The exploitation of this flexibility allows us to adapt the solution process to the *SD-BB* algorithm. Therefore, basic procedures such as branching, computing lower bounds, generating feasible solutions, pruning, and checking feasibility have to be overwritten with our developed procedures.

All computations were performed on an AMD 64 X2 Dual Core 4800+ platform with 2.4 Ghz and 4 GB of main memory.

8.1 Test Instances

For the computational investigations, we consider various test instances of the *OPGen* problem introduced in Chapter 3. A test instance of the problem is described by a combination of sets and parameters which characterize the power generation system together with the consumers' load, the available wind power and the electricity prices. In Section 8.1.1, each facility type considered in the power generation system is specified defining the first part of an instance. The second part consists of the data representing the available wind power, the electricity prices, and the consumers' demand. As explained in Section 3.2, in case of the *S-OPGen* problem, uncertainty in electricity prices and in available wind power is included in the model. Therefore, in Section 8.1.2 the generation of the corresponding scenario trees is discussed.

8.1.1 Facilities in the Power Generation System

As discussed in Section 2.3, we restrict our consideration to the application of selected facilities within the power generation system. More precisely, the system consists of coal power plants in order to cover the base load, fast gas turbine power plants capable of reacting to short term fluctuations together with both types of energy storages, i.e., pumped hydro storages (PSW) and compressed air energy storages (CAES). Below, we highlight the basic characteristics of the facilities focusing on the consequences for the generic model described in Chapter 3. To this end, we specify the most important parameters appearing in the corresponding constraints presented in Section 3.1.3. We remark that all operating parameters of these facilities rely on real world data which are obtained from [Epe07] and [Gri07].

Coal Power Plants

At first, we consider a typical power plant $i \in \mathcal{I}$ which is based on hard coal. For this type of thermal power plant, frequent start-up and shut down processes should be avoided in order to reduce thermal stress. Hence, a minimum running time θ_i^{up} and minimum down time θ_i^{down} is assumed, i.e., $\theta_i^{up}, \theta_i^{down} > 1$. Further important characteristics are the maximal power gradient $\Delta_i^{max} < p_i^{max} - p_i^{min}$ and the start-up costs $\gamma_i^{up} > 0$ appearing in the objective function. Recall that p_i^{min} and p_i^{max} represent the minimum and maximum production of a power plant i. Finally, we remark that a coal power plant shows a relatively high efficiency $\eta_i(p_i)$ in comparison to a gas turbine power plant.

Gas Turbine Power Plants

This type of power plant provides the ability to be turned on and off within minutes. Hence, minimum running time and down time restrictions can be neglected in this case, which means $\theta_i^{up} = 1$ and $\theta_i^{down} = 1$. For the same reason, the power gradient Δ_i^{max} also can be disregarded. However, as in case of the coal power plant, start-up costs $\gamma_i^{up} > 0$ are considered in the objective function. As mentioned above, the gas turbine plant exhibits a lower efficiency than the coal power plant and in case of partial load the efficiency strongly decreases. Nevertheless, a gas turbine is ideally suited for covering peak load due to its flexibility.

Pumped Hydro Storages

The data for this facility is based on the PSW located in Geesthacht, Germany. In the storage, three sets of gas turbines and pumps are operated, where the pumps are responsible for charging and the turbines are used to produce electricity. Each pair is connected to a water reservoir of higher elevation by a pipe. Consequently, the cardinality of the set of charging units $\mathcal{K}_j$ and discharging units $\mathcal{L}_j$ of a PSW for $j \in \mathcal{J}$ equals three, i.e., $|\mathcal{K}_j| = 3$ and $|\mathcal{L}_j| = 3$. Due to technical characteristics, the start-up energy α_j^{in} as well as the start-up costs $\gamma_j^{in,up}$ of a charging unit are negligibly small and we set $\alpha_j^{in} = 0$ and $\gamma_j^{in,up} = 0$. For the same reasons, the start-up energy α_j^{out} and the start-up costs $\gamma_j^{out,up}$ resulting from a discharging unit are neglected yielding $\alpha_j^{out} = 0$ and $\gamma_j^{out,up} = 0$. Due to the high efficiency of the charging and discharging units, the overall efficiency of this type of storage is rather high in comparison to the compressed air energy storage described in the following paragraph.

Compressed Air Energy Storages

Next to the PSW, we also include a CAES in our energy system. The data used for our computations is based on the storage in Huntdorf, Germany, which is the first one worldwide. The CAES basically consists of a compressor unit and a gas turbine which are responsible for charging and discharging, respectively, leading to $|\mathcal{K}_j| = 1$ and $|\mathcal{L}_j| = 1$ for $j \in \mathcal{J}$. In contrast to the PSW, the CAES requires start-up energy for charging and discharging, i.e., $\alpha_j^{in}, \alpha_j^{out} > 0$, which are considered in the storage balance equation (3.12) described in Section 3.1.3. Furthermore, the turbine integrated in the

CAES consumes additional gas which results in costs $\gamma_j^{out,up} > 0$ considered in the objective function. However, start-up costs of the compressor unit are negligible and hence, we set $\gamma_j^{in,up} = 0$.

Further information on all types of facilities is given in Section 2.3.1. Altogether, the underlying energy system of a problem instance consists of various facilities of the plants and storages describe above. In particular, facilities of the same type are assumed to show the same technical characteristics and dimensioning. Consequently, the energy system can be characterized by a tuple

$$s = (n_C, n_G, n_P, n_A),$$

where n_C represents the number of coal power plants, n_G the number of gas turbine power plants, n_P the number of pumped hydro storages, and n_A the number of compressed air energy storages.

8.1.2 Stochastic Data

As indicated above, the uncertainty considered in the model concerns the wind power and the price for electricity which is modeled via a scenario tree. For their generation, sets of 1000 initial scenarios are used, which include quarter-hourly values for the wind power and the electricity price taking their correlation into account. These data are provided by [Wol08] applying the following approach.

A time series model is used to describe the stochastic wind power process which is adjusted to historical data. The expected spot market prices are derived from a fundamental model which is based among other factors on the current power plants in Germany, the wind process described above and prices for fuel to account for the interdependency of the wind power and the electricity price behavior. The fluctuations of the spot market prices are achieved by using a further time series model. For more information, we refer to [EKR+09] and [Web06].

Due to the computational complexity resulting from these scenarios, a scenario reduction algorithm is applied which selects a subset of the scenarios and assigns appropriate probabilities to those maintained. To this end, we make use of the tool GAMS/SCENRED which approximates the full scenario tree by a smaller one which exhibits a reduced number of scenarios by using the algorithm presented in [DGKR03].

Table 8.1: Test instances for parameter tuning

	Facilities				Scenario Tree			
Instance	CP	GTP	PSW	CAES	T	N	S	Load
tuneInst1	1	1	1	1	48	144	13	w
tuneInst2	1	1	1	1	60	90	3	w
tuneInst3	1	1	1	1	60	90	3	s
tuneInst4	1	1	1	1	96	371	13	w
tuneInst5	1	1	1	1	96	823	34	w
tuneInst6	1	1	5	5	60	90	3	w
tuneInst7	5	5	1	1	60	90	3	w
tuneInst8	5	5	5	5	60	90	3	w

8.1.3 Test Instances for Parameter Tuning

For a good performance of the *SD-BB* algorithm, a tuning of the involved parameters applied to a given problem is essential. Since we focus on the solution of the *S-OPGen* problem, in the following we specify eight test instances of this problem which provide the basis for the parameter calibration. In order to produce meaningful results, we consider test instances which exhibit different characteristics of the problem, i.e., varying in the energy system, the underlying scenario tree, and the consumers' demand. The selected eight test instances are summarized in Table 8.1.

In the first column, the name of the instance is stated. The second through fifth columns describe the tuple defining the energy production system under consideration. The following columns characterize the underlying scenario tree. In detail, the column labeled by "T" shows the number of time steps of the scenario tree, while the columns denoted by "N" and "S" represent the number of nodes and the number of scenarios of the corresponding scenario tree, respectively. The last column specifies the consumers' demand, where the letter "w" indicates that the corresponding data is derived from a typical winter week while "s" refers to a summer week.

Regarding the energy system, the first five instances coincide completely. For tuneInst6 and tuneInst7, we increase the number of power plants and energy storages, respectively, in order to be able to evaluate their impact on the performance of the solution process independently. In the last instance, the number of all types of facilities is augmented. In summary, these instances show variations in the basic properties characterizing an instance of the *S-OPGen* problem. Hence, we assume that this selection provides a reliable basis for the parameter calibration of the methods applied within the *SD-BB* framework.

8.2 Separation

In this section, we investigate the effect of adding the constraints describing the stochastic switching polytope of Chapter 4 as cutting planes to a branch-and-cut algorithm. To be more precise, we consider the facets which are derived from the complete linear description of the minimum runtime and downtime conditions, as introduced in Section 4.3. Within our computational investigations, we compare the following three approaches of modeling these additional restrictions for the dynamic behavior of power plants: At first, we add the original constraints (4.2) through (4.5) explicitly to the problem formulation. In a second approach, we extend this modeling by applying additionally the separation algorithm of inequalities (4.10) through (4.12) in each node of the branch-and-cut algorithm. Finally, the original constraints are completely omitted from the problem formulation and are imposed implicitly using the corresponding separation methods.

In order to evaluate these approaches, we perform a series of test runs solving eight test instances which are based on two different scenario trees. In detail, the instance instSep1 through instSep4 rely on the same scenario tree as tuning instance tuneInst5 specified in the previous section, where the time horizon under consideration varies from 48 to 96 time steps. The instances instSep5 through instSep8 make use of the scenario tree obtained from tuneInst4 showing stronger fluctuations in the available wind power. For the computations, we fix the minimum runtime and downtime of the coal power plants to five hours, respectively, which constitute realistic time spans based on real world operations, see e.g. [Gri07]. Formally, we set $\theta^{up} = 20$ and $\theta^{down} = 20$ for all coal power plants assuming a time discretization of 15 minutes. Additionally, we impose a time limit of 3600 CPU seconds for the running time of the solution process.

At first, we assume the start-up costs to equal zero in order to allow a more variable behavior of the plants, i.e., $\gamma^{up} = 0$. Although our original model takes start-up cost of the coal power plants into account, this approach is chosen to facilitate variations of the operational level of the power plant. This provides the basis for the application of the cutting planes obtained from the switching polytope. Subsequently, we also investigate variations of the start-up costs. Tables 8.2, 8.3, and 8.4 summarize the computational

Table 8.2: Computational results based on incorporating the original minimum runtime and downtime restrictions explicitly

Instance	T	# Con.	# Nodes	Lower Bd.	Upper Bd.	Time	Gap %
instSep1	48	10537	2755	116718.2	116729.3	148.3	0
instSep2	60	18121	8717	162936.3	162952.2	817.5	0
instSep3	72	29039	16344	217880.9	218089.9	3600.0	0.10
instSep4	96	75809	2095	348927.1	360989.1	3600.0	3.34
instSep5	48	9111	9094	673553.3	673620.25	172.8	0
instSep6	60	14013	31663	1073300.4	1073400.4	718.5	0
instSep7	72	25093	89486	1561800.5	1562190.0	3600.0	0.02
instSep8	96	102699	11210	2559310.8	2560460.9	3600.0	0.05

Table 8.3: Separating minimum runtime and downtime restrictions combined with modeling the original constraints explicitly

Instance	T	# Con.	# Nodes	# Cuts	Lower Bd.	Upper Bd.	Time	Gap %
instSep1	48	10537	1568	1	116721.4	116729.3	104.2	0
instSep2	60	18121	8600	2	162946.2	162952.4	778.7	0
instSep3	72	29039	16124	2	217770.9	218106.4	3600.0	0.15
instSep4	96	75809	6690	2	349108.0	356892.7	3600.0	2.18
instSep5	48	9111	8274	3	673552.3	673620.1	144.9	0
instSep6	60	14013	17798	4	1073310.25	1073410.0	459.7	0
instSep7	72	25093	73800	3	1561850.5	1562180.5	3600.0	0.02
instSep8	96	102699	10982	4	2559319.8	2560459.7	3600.0	0.04

Table 8.4: Separating minimum runtime and downtime restrictions omitting the original constraints

Instance	T	# Con.	# Nodes	# Cuts	Lower Bd.	Upper Bd.	Time	Gap %
instSep1	48	8227	2937	16	116721.7	116729.3	87.8	0
instSep2	60	13939	5920	22	162946.1	162955.1	298.2	0
instSep3	72	22166	47226	21	217821.1	218073.2	3600.0	0.12
instSep4	96	57392	11101	22	356326.1	348741.3	3600.0	2.14
instSep5	48	7152	26564	48	673552.2	673620.6	285.0	0
instSep6	60	10845	16703	60	1073280.1	1073380.2	345.5	0
instSep7	72	19195	71496	137	1561820.25	1562170.1	3600.0	0.02
instSep8	96	77652	15431	444	2559020.1	2560520.7	3600.0	0.06

results obtained by applying one of the three modeling versions to describe the minimum runtime and downtime restrictions. Besides the name of the considered instance stated in the first column of Table 8.2, the second column shows the number of time steps ranging from 48 to 96 for both problems. The third column gives the number of constraints of the instance at hand which indicates the variation in problem size for the different modeling approaches. The number of nodes of the branch-and-bound tree is presented in the next column. While the column "Lower Bd." describes the best value of the LP relaxation found within the time limit of 3600 CPU seconds, the value of the best feasible solution is shown in column "Upper Bd.". The column before last documents the running time in CPU seconds which is smaller than 3600 if an optimal solution is found before the time limit is reached. The last column presents the relative difference between the best lower and the upper bound. In the following we assume a solution to be optimal when the relative gap is less than 0.01 %, which we indicate by a "0". In contrast to the first table, Tables 8.3 and 8.4 contain an additional column denoted by "# Cuts", where the number of cuts separated during the solution process is shown.

Comparing the results of the first two tables, in most of the cases we observe a decrease in the number of branch-and-bound nodes by additionally applying the separation algorithm. Although only few constraints are added during the solution process, the lower bound is increased significantly with respect to the large instances of 72 and 96 time steps. The improved relative gap becomes particularly apparent for instSep4. Additionally, we detect a weak trend of separating more inequalities for the instances relying on strongly fluctuating wind power, i.e., instSep5 to instSep8, than for instances based on a more regular power supply. We believe that this effect is caused by the growing regulative tasks of the power plant in case of strongly varying wind supply which possibly results in a violation of the switching constraints. Regarding the running time of the instances solved to optimality, the application of the cutting plane approach yields a reduction for all instances considered.

Taking Table 8.4 into account, the sizes of the current instances are significantly reduced due to the neglect of the original minimum runtime and downtime restrictions in the problem formulation. More precisely, the number of constraints is downsized by a factor of approximately 0.8. As a consequence, the number of cuts added during the solution process increases considerably. Comparing the running time spent for the solution of the smaller instances of 48 and 60 time steps with the results of the other tables, in

Table 8.5: Separation of the minimum runtime and downtime conditions comparing different start-up costs

Method	γ^{up}	# Con.	# Nodes	# Cuts	Lower Bd.	Upper Bd.	Time	Gap %
explicit	0	14013	31663	-	1073295.5	1073403.5	756.7	0
	5	14013	22485	-	1073266.1	1073367.5	534.7	0
	10	14013	14722	-	1073284.4	1073392.6	547.0	0
	40	14013	23083	-	1073289.5	1073398.0	466.4	0
explicit	0	14013	16770	1	1073288.8	1073396.8	434.6	0
& separ.	5	14013	23430	1	1073278.4	1073386.5	574.3	0
	10	14013	32374	2	1073258.7	1073367.2	723.8	0
	40	14013	18898	0	1073280.1	1073388.0	445.4	0
separ.	0	10845	16703	60	1073288.8	1073396.8	327.2	0
	5	10845	22460	48	1073280.9	1073380.9	346.6	0
	10	10845	28003	39	1073299.4	1073399.8	548.4	0
	40	10845	18069	3	1073260.6	1073370.2	348.1	0

three of four cases this approach shows the best performance terminating significantly faster. This observation is emphasized by the small relative differences of the lower and upper bounds which are obtained for larger planning horizons in Table 8.4. Considering these results, we can conclude that the separation algorithm without formulating the original constraints improves the solution process of the branch-and-cut framework. We remark that this suggestion is derived under the assumption of omitting the start-up costs of the power plants. Hence, in the following we extend our computational investigations by the incorporation of start-up costs.

For the evaluation of the separation approach under the consideration of start-up costs, we exemplarily focus on the solution process of tuneInst6 for different values of the costs γ^{up}. Within the computational studies, we set γ^{up} to values ranging from 0 to 40 euros per MWh since the latter parameter constitutes a realistic value to assess the costs evolving from starting up a coal power plant, compare e.g. [Gri07]. The results of the test runs are shown in Table 8.5, where the first column states the chosen modeling approach of the minimum runtime and downtime restrictions, as described above. Additionally, the second column indicates the chosen value for the start-up costs γ^{up}. The remaining columns are denoted analogously to the previous tables.

As expected, the number of separated constraints decreases with higher start-up costs, since they prevent the power plants from turning off in times of low demand. Nevertheless, the results emphasize the observation made for the previous three tables, which suggests the application of the separation algorithm by completely omitting the explicit modeling. Indeed, the strong decrease in size results in less computational effort to solve the LP relaxation in the branch-and-bound nodes. Together with the efficient separation algorithm, this combination yields the best performance.

8.3 Heuristics

Besides the improvement of the lower bound, we have the possibility of generating good feasible solutions for the *D-OPGen* and *S-OPGen* problem by applying an approximate-and-fix approach, as described in Chapter 5. More precisely, we can use the rolling horizon procedure presented in Section 5.2 in order to determine a feasible solution of the deterministic *D-OPGen* problem, while the approximate-and-fix algorithm of Section 5.3 serves as a heuristic for the *S-OPGen* problem. Based on the outcome of a series of test runs, we determine a set of suitable combinations of parameter values for both methods applied to the problem.

8.3.1 Rolling Horizon Algorithm

In this section, we consider the applicability of the rolling horizon heuristic for the solution of the *D-OPGen* problem. We start by tuning the corresponding parameters of the heuristic based on a series of test instances in analogy to the stochastic test set described in Section 8.1.3. Subsequently, we evaluate its performance for larger instances.

For the parameter calibration, we use a set of six test instances which are summarized in Table 8.6. The first six columns are denoted in analogy to Table 8.1. Column seven named "Data" represents the input values corresponding to the available wind power, the consumers' demand, and the prices for electricity. In detail, the letter "w" indicates that the values rely on historical data of a typical winter day while "s" refers to the data of a typical summer day. Finally, the last three columns specify the size of the problem at hand. Note that instance tuneInsDet1 through tuneInstDet3 vary in the number of time steps and in the input data while the last three instances show variations in the power generation system under consideration.

Table 8.6: Test instances for the parameter tuning of the rolling horizon heuristic

Instance	CP	GTP	PSW	CAES	T	Data	# Var.	# Bin.	# Con.
tuneInstDet1	1	1	1	1	96	w	12296	5380	11337
tuneInstDet2	1	1	1	1	144	s	18440	8068	17001
tuneInstDet3	1	1	1	1	192	s	24584	10756	22665
tuneInstDet4	5	5	1	1	48	w	11156	5004	10481
tuneInstDet5	1	1	5	5	48	w	25560	11148	23365
tuneInstDet6	5	5	5	5	48	w	30564	13460	28173

In order to be able to evaluate the quality of the solution found by the rolling horizon heuristic, CPLEX is employed to compute the optimal objective function value of the current instance. Remark that we applied a time limit of 3600 CPU for the solution process. In Table 8.7, the computational results obtained by CPLEX concerning the six instances presented above are summarized. The first column specifies the test instance, while the number of branch-and-bound nodes as well as the best value of the LP relaxation found within the time limit of 3600 CPU seconds are presented in the second and third column, respectively. The value of the best feasible solution and the corresponding CPU time are presented in the column "Upper Bd." and "Time". In the last column, the relative difference between the lower and upper bound is computed.

We observe that after the expiration of 3600 seconds, half of the instances have been solved to optimality. For tuneInstDet3, tuneInstDet5, and tuneInstDet6, we only obtain a lower bound on the optimal objective function value and thus, we utilize this value for evaluating the quality of the solution computed by the rolling horizon heuristic for these instances.

For the solution of the mixed-integer subproblems occurring in each iteration of the rolling horizon algorithm, we also make use of the solver CPLEX. Concerning the default setting of the rolling horizon heuristic, we choose approximation strategy $\tilde{S}_1$ for the following computations which reformulates the problem associated with the approximated period as described in Section 5.2.1. Furthermore, we limit the running time of each subproblem to t_{max} whose value is determined problem specifically. More precisely, we set the scaling factor k which defines t_{max} to $k = 10$, which is described in detail later in this section. On this basis the parameter tuning of the heuristic is performed, starting with the consideration of the basic parameters defining the size of the subproblem.

Table 8.7: Computational results using CPLEX with a time limit of 3600 sec.

Instance	# Nodes	Lower Bd.	Upper Bd.	Time	Gap %
tuneInstDet1	8141	1879678.3	1879867.2	166.1	0
tuneInstDet2	8856	1623171.9	1623336.7	355.1	0
tuneInstDet3	90829	2317355.6	2317981.8	3600.0	0.03
tuneInstDet4	59467	344700.6	344735.3	1636.8	0
tuneInstDet5	34421	927410.7	932163.7	3600.0	0.50
tuneInstDet6	12474	342864.1	342990.8	3600.0	0.04

Selection of Parameters T^{ex}, T^{shift}, and T^{app}

With the aim of calibrating the rolling horizon heuristic, at first we concentrate on determining a general setting of the parameters T^{ex}, T^{shift}, and T^{app}, as they define the subproblems which are iteratively solved during the solution process. Recall that T^{ex} describes the number of time steps which are modeled exactly and T^{shift} indicates by how many steps the exact phase is shifted after each iteration. The number of time steps which are approximated in a subproblem is denoted by T^{app}, as explained in Section 5.2. The performance of different combinations is measured by the quality of the solution, i.e., by the gap to the optimal objective function value and the running time of the process. The according numerical results are shown in Table 8.8.

In the columns of the table which are labeled by T^{shit}, T^{ex}, and T^{app}, the values of the corresponding parameters are varied depending on the number of time steps of the instance at hand. A "*" appearing in column T^{app} indicates that in each iteration of the heuristic, the approximated period comprises all time steps up to T. Remember that otherwise only T^{app} time steps are approximated and the remaining ones are completely neglected. The column "# Iter." shows the number of iterations of the rolling horizon algorithm which results from the values listed in the three previous columns. The objective function value of the best feasible solution found by the algorithm is shown in column "Upper Bd.", while the CPU time spent for its computation is given in column "Time". Finally, the relative difference between this solution value and the optimal objective function value or the best lower bound found is presented making use of the results which are shown in Table 8.7.

At first, we consider instances tuneInstDet1 to tuneInstDet3. We observe that for all combinations of the parameters T^{shift}, T^{ex}, and T^{app} the heuris-

Table 8.8: Computational results of the rolling horizon heuristic for determining T^{shift}, T^{ex}, and T^{app}

Instance	T^{shift}	T^{ex}	T^{app}	# Iter.	Upper Bd.	Time	Gap %
tuneInstDet1	12	24	24	7	1881248.7	22.7	0.08
	12	24	*	7	1880247.7	40.2	0.03
	24	48	*	3	1879995.8	42.3	0.02
tuneInstDet2	12	24	72	11	1623708.8	63.3	0.03
	24	48	48	5	1623241.6	57.3	0
	24	48	*	5	1623676.1	67.1	0.03
tuneInstDet3	24	48	48	7	2322073.5	77.2	0.20
	24	48	96	7	2324278.0	87.2	0.250
	24	48	*	7	2321577.7	96.2	0.18
tuneInstDet4	6	12	12	7	365238.2	28.4	5.62
	6	12	*	7	344780.0	54.3	0.02
	12	24	*	3	345046.1	52.6	0.10
tuneInstDet5	6	12	12	7	1114217.7	53.5	16.77
	6	12	*	7	947837.3	79.0	2.16
	12	24	*	3	933930.2	71.2	0.70
tuneInstDet6	6	12	12	7	352433.1	102.2	2.68
	6	12	*	7	343044.4	135.7	0.05
	12	24	*	3	343643.8	119.1	0.19

tic performs remarkably well with respect to the solution quality, i.e., the gap is always less than 0.25 %. As expected, we observe a slight increase in the solution time if the complete planning horizon is approximated which is indicated by a "*" in column T^{app}. However, for this setting the heuristic finds feasible solutions of highest quality. Additionally, we note that for the first three tuning instances a combination of $T^{ex} = 48$ and $T^{shift} = 24$ yields the best results. However this combination performs worse for the last instances. Recall that for tuneInst4 through tuneInst6 a planning horizon of 48 time steps is considered. Therefore, this setting would yield a solution in one iteration as the exact period would comprise the entire planning horizon. Thus, $T^{ex} = 48$ and $T^{shift} = 24$ is inappropriate for these instances. In contrast, the values $T^{ex} = 24$ and $T^{shift} = 12$ seem to be a better choice. This different outcome may be explained by the small number of facilities of the first instances resulting in subproblems of smaller size, whereas the energy systems of the last three instances comprise 12 to 20 facilities, compare Table 8.6, yielding subproblems which are harder to solve.

Altogether, we deduce the following trend for a suitable setting of the parameters providing the basis for the default values of the heuristic: For test instances with less than five facilities, we set $T^{shift} = 24$, $T^{ex} = 48$, and $T^{app} = 48$, while for larger systems we specify the values $T^{shift} = 12$, $T^{ex} = 24$, and $T^{app} = 72$. In both cases, this means that we use a foresight of one day, providing sufficient information for almost optimal decisions in the exact period. In the following, we select values for T^{shift}, T^{ex}, and T^{app} according to the specification above.

Selection of the Approximation Strategy

Having determined the basic construction parameters for the subproblems, in the next step we investigate the impact of the different approximation methods $\bar{S}_1, \tilde{S}_1$, and S_2 presented in Section 5.2.1 on the performance of the heuristic. Recall that the first two strategies both approximate the piecewise linear efficiency functions by linear functions. More precisely, $\tilde{S}_1$ yields a closer approximation than $\bar{S}_1$, as it includes an additional constant term. On the other hand it also involves an additional binary variable in the formulation. Strategy S_2 yields a coarsening of the approximated period by aggregating a certain number of time steps to one.

The computational results are shown in Table 8.9, where the second column "Approx." reflects the chosen approximation strategy. In case strategy S_2 is chosen, the column labeled by "Agg." indicates how many time steps are aggregated to one step. The remaining columns are denoted analogously to Table 8.8.

Comparing the quality of the solutions applying these strategies, we detect a clear dominance of strategies $\bar{S}_1$ and $\tilde{S}_1$. Obviously, the aggregation of time steps results in a significant loss of information yielding solutions with higher objective function values. But as expected, the application of strategy S_2 clearly decreases the running time, since the corresponding subproblems are reduced in size. Regarding the quality as well as the running time under the application of strategy $\bar{S}_1$ and $\tilde{S}_1$, the results are almost identical. Nevertheless, for the instances consisting of energy systems of larger size, $\tilde{S}_1$ shows a slightly better performance which is why we choose $\tilde{S}_1$ as default approximation strategy.

Table 8.9: Computational results of the rolling horizon heuristic for comparing different approximation strategies

Instance	Approx.	Agg.	# Iter.	Upper Bd.	Time	Gap %
tuneInstDet1	$\bar{S}_1$	-	3	1879940.8	37.7	0.01
	$\tilde{S}_1$	-	3	1879995.8	41.7	0.02
	S_2	2	3	1948888.6	11.2	3.55
	S_2	4	3	1953449.8	5.3	3.78
tuneInstDet2	$\bar{S}_1$	-	5	1623308.0	50.9	0
	$\tilde{S}_1$	-	5	1624574.7	53.6	0.06
	S_2	2	5	1845235.3	18.4	12.03
	S_2	4	5	1883827.2	15.2	13.09
tuneInstDet3	$\bar{S}_1$	-	7	2320105.5	75.7	0.12
	$\tilde{S}_1$	-	7	2322073.5	77.0	0.20
	S_2	2	7	2586817.9	24.5	10.42
	S_2	4	7	2666530.9	17.4	13.42
tuneInstDet4	$\bar{S}_1$	-	3	345303.0	54.6	0.17
	$\tilde{S}_1$	-	3	345046.1	32.6	0.10
	S_2	2	3	359351.8	29.8	4.08
	S_2	4	3	359100.0	31.0	4.01
tuneInstDet5	$\bar{S}_1$	-	3	932311.6	70.4	0.53
	$\tilde{S}_1$	-	3	933930.2	70.2	0.70
	S_2	2	3	1060764.8	31.5	12.57
	S_2	4	3	1033929.2	32.1	10.250
tuneInstDet6	$\bar{S}_1$	-	3	344383.8	118.0	0.44
	$\tilde{S}_1$	-	3	343671.6	118.8	0.23
	S_2	2	3	406382.0	97.6	15.63
	S_2	4	3	401277.9	97.9	14.56

Limiting the Running Time of the Subproblems

As mentioned above, the solution time of the subproblems is restricted to a prespecified value aiming at reducing the overall running time. This approach is motivated by the observation that even though the subproblems are not solved to optimality, solutions of high quality can be obtained. However, the time limit has to be chosen carefully in order to avoid a significant loss of quality.

Table 8.10: Computational results of the rolling horizon heuristic for determining the time scaling factor k

Instance	k	# Iter.	Upper Bd.	Time	Gap %
tuneInstDet1	1	3	1879954.5	131.2	0.01
	10	3	1879995.8	42.3	0
	50	3	1880745.9	10.4	0.06
tuneInstDet2	1	5	1623224.5	436.4	0
	10	5	1623708.8	63.3	0
	50	5	1629459.1	16.1	0.259
tuneInstDet3	1	7	2318034.0	510.9	0.03
	10	7	2322073.5	77.2	0.14
	50	7	2325018.0	18.5	0.253
tuneInstDet4	1	3	344777.2	239.3	0.02
	10	3	345046.1	32.6	0
	50	3	346368.5	13.2	0.48
tuneInstDet5	1	3	931845.1	596.8	0.48
	10	3	933930.2	71.2	0
	50	3	1154377.4	30.6	19.66
tuneInstDet6	1	3	342978.1	755.4	0.04
	10	3	343643.8	119.1	0.19
	50	3	359864.5	39.9	4.72

Obviously, an adequate choice of the time limit depends on the specific problem whose size is strongly influenced by the number of time steps T and the underlying energy system, i.e., the number of facilities. Hence, we specify a time limit t_{max} in dependence of both aspects. Additionally, we account for the number of iterations of the heuristic which also provides an indication of the problem size. Since the complexity of the problems increases with an augmented number of time steps and facilities while a larger number R of iterations most likely yields a reduction in problem size, we set

$$t_{max} = \frac{T(n_C + n_G + n_P + n_A)}{k\,R}, \tag{8.1}$$

where n_C, n_G, n_P, and n_A represent the number of coal power plants, gas turbine power plants, PSWs, and CAESs, respectively. Additionally, we involve a parameter $k \in \mathbb{R}_+$ for scaling the solution time, which is chosen based on a series of test runs whose results are summarized in Table 8.10. The denotation of the table is carried out as in the Table 8.9 except for the second column which represents the current time scaling factor k used for the computation of t_{max}. For the test runs, we set k to 1, 10, and 50 yielding a moderate to strong time limit t_{max}.

As intended, we observe a clear reduction of the running time if we augment k. Furthermore, the quality of the solutions decreases with increasing k, as expected. Within the scaling factors considered, the value $k = 10$ performs best, as we obtain solutions showing a gap of less than 0.2 % for all test instances at hand. Since additionally the computational time is significantly reduced in comparison to $k = 1$, we choose $k = 10$ as default value.

Applying the Rolling Horizon Heuristic to Large Instances

Finally, we evaluate the performance of the heuristic based on a set of large instances applying the default parameter setting which has been selected in the previous sections. In summary, we fix the following values and methods:

- In case the number of facilities is less than five, we set $T^{shift} = 24$, $T^{ex} = 48$, and $T^{app} = 48$, otherwise $T^{shift} = 12, T^{ex} = 24$, and $T^{app} = 72$.
- We apply the approximation strategy $\tilde{S}_1$.
- We set the scaling factor k to 10.

We consider ten instances of increased size varying in the length of the planning horizon and in the number of facilities within the underlying energy generation system. To be more precise, we consider planning horizons with up to 480 time steps which corresponds to five days and instances with a generation system of up to 32 facilities. In order to measure the quality of the solutions generated by the rolling horizon heuristic and to indicate the computational effort to solve the instances to optimality, we additionally solve the problems by CPLEX which is able to provide an optimal objective function value or at least a lower bound for this value. For these larger instances, we restrict the running time to 10000 CPU seconds. Table 8.11 summarizes the numerical results obtained from the rolling horizon heuristic and CPLEX.

For a specification of the test instance, the number T of time steps of the planning horizon is stated within the name of the instance which is shown in

Table 8.11: Computational results of the rolling horizon heuristic applied to large instances

Instance	System	# Var.	# Bin.	# Con.	Meth.	Upper Bd.	Time	Gap %
instD1_96	(1111)	12296	5380	11337	*RH*	1879995.7	41.6	0.01
					Ex	1879678.3	177.9	0
instD2_144	(1111)	18440	8068	17001	*RH*	3418850.8	48.1	0.02
					Ex	3418081.7	1661.2	0
instD3_288	(1111)	36872	16132	33993	*RH*	5298995.8	109.3	0.10
					Ex	5293503.3	10000.0	0.01
instD4_384	(1111)	49160	21508	45321	*RH*	6381197.5	145.53	0.47
					Ex	6357532.5	10000.0	0.10
instD5_480	(1111)	61448	26884	56649	*RH*	10494049.7	184.5	0.28
					Ex	10469643.5	10000.0	0.05
instD6_96	(3333)	36694	16140	33819	*RH*	770346.4	146.8	2.31
					Ex	752374.2	10000.0	0.11
instD7_96	(5555)	61092	26900	56301	*RH*	753180.25	401.7	0.13
					Ex	752374.2	10000.0	0.03
instD8_96	(5588)	90192	39578	82818	*RH*	3091892.0	521.2	3.02
					Ex	3003807.9	10000.0	0.18
instD9_96	(8855)	68589	30362	63507	*RH*	2306084.95	373.7	0.02
					Ex	2305503.2	10000.0	0.01
instD10_96	(8888)	97689	43040	90024	*RH*	2308830.73	646.04	0.09
					Ex	2307723.0	10000.0	0.05

the first column of the table. Additionally, the column denoted by "System" describes the number of facilities included in the generation system. Recall that the tuple (n_C, n_P, n_P, n_A) represents the number of coal power plants, gas turbine power plants, PSWs, and CAESs. The third to fifth column reflect the size of the resulting MIP, listing the total number of variables, the number of binary variables and the number of constraints. As mentioned above, we compare our rolling horizon heuristic (*RH*) with the exact solver CPLEX (*Ex*) which is indicated in the column "Meth.". In analogy to the previous tables, the last three columns comprise the objective function value of the best feasible solution available after 10000 CPU seconds together with the running time and the relative gap. In case of the rolling horizon heuristic, the gap is computed based on the bounds obtained from CPLEX. We remark that the running times of the heuristic and of CPLEX are hardly comparable as the latter one relies on an exact solution algorithm providing a quality certificate of the generated solutions. Nevertheless, the running times spent by CPLEX demonstrate the significant increase in computational effort being necessary to solve the instances to optimality when the considered planning horizons are enlarged.

Regarding the first five instances, the rolling horizon heuristic performs well with respect to the quality of the solutions which are determined in less than three minutes. All these solutions show a relative gap of less than 0.5 %. The consideration of the running time of these instances demonstrates the great advantage of this approach: Having fixed the approximated period to a predefined number of time steps, the consumed running time increases only linearly with the total number of time steps in the rolling horizon. For the last test instances, this approach shows a lower quality of the solutions when more facilities are integrated in the generation system, particularly apparent for InstD6_96 and instD8_96. However, for the last two instances the heuristic generates almost optimal solutions.

Summarizing the results, we conclude that the heuristic is well suited for the applications to test instances with large planning horizons. Since the running time only increases linearly with the considered time steps, the heuristic provides the possibility of generating near optimal solutions in relatively short running times. Motivated by these good results, this approach is transferred to the stochastic case and the resulting approximate-and-fix heuristic is computationally investigated in the following section.

8.3.2 Approximate-and-Fix Heuristic

For the determination of a good feasible solution of the *S-OPGen* problem, we have the possibility of applying the approximate-and-fix heuristic presented in Section 5.3. The aim of this section is the evaluation of the proposed method yielding a parameter setting suitable for our purposes. Therefore, we execute a series of test runs based on the tuning instances presented in Section 8.1.3 which reflect several variations of the characteristics of the problem at hand.

Motivated by the close relation between the approximate-and-fix heuristic and the rolling horizon method together with the affinity of the *S-OPGen*

Table 8.12: Computational results using CPLEX with a time limit of 3600 sec.

Instance	# Nodes	Lower Bd.	Upper Bd.	Time	Gap %
tuneInst1	176529	572153.4	572153.4	3600.0	0.05
tuneInst2	287325	281118.3	281159.7	3600.0	0.01
tuneInst3	6388	169906.4	169923.1	112.3	0
tuneInst4	62441	1782634.5	1783512.4	3600.0	0.05
tuneInst5	6546	625823.0	625662.5	3600.0	0.03
tuneInst6	10978	1878825.5	1882847.5	3600.0	0.21
tuneInst7	130717	1783178.6	1783750.2	3600.0	0.03
tuneInst8	1414	495041.9	498016.1	3600.0	0.59

and *D-OPGen* problem, we have decided to transfer selected results of the previous section to the present one. To be more precise, we apply the approximation strategy $\tilde{S}_1$ within the approximate-and-fix heuristic due to the computational results shown in Table 8.9. Furthermore, we approximate the entire planning horizon, since the results of Table 8.8 indicate that this setting yields good quality solutions for the planning horizon we consider here. We remark that for large planning horizons a restriction of the approximated period may be reasonable. However, the application of approximation method $\tilde{S}_1$ allows the consideration of the entire horizon, as it strongly reduces the problem size yielding considerably good results for all instances investigated. Consequently, the parameter tuning is restricted to the two parameters T^{ex} and T^{shift} and the scaling factor k which determines the time limit of the subproblems.

Since we aim at evaluating the approximate-and-fix heuristic based on the quality of the generated feasible solutions, we solve the tuning instances tuneInst1 through tuneInst8 by CPLEX in order to obtain an optimal solution value in analogy to the deterministic case. Again we specify a time limit of 3600 CPU seconds. The numerical results are shown in Table 8.12 where the notation of the columns is inherited from Table 8.7. Although all instances show a planning horizon of less than 96 time steps, compare Table 8.1, only tuneInst3 is solved to optimality within the time limit. Nevertheless, the relative gap between the best lower bound and best upper bound is less than 0.6 % for all instances providing a reliable basis for the evaluation of the approximate-and-fix heuristic.

Selection of Parameters T^{shift} and T^{app}

We start our computations by comparing several combination of the parameters T^{ex} and T^{shift} based on the solution of the eight instances tuneInst1 through tuneInst8. Besides the setting of the parameters described above, we also restrict the running time allowed for the subproblem according to (8.2) by setting $k = 50$ as default value. The results are summarized in Table 8.13 using an identical notation of the columns as in the previous section except for the fifth column. Here, the column "# Iter." contains the total number of iterations of the approximate-and-fix heuristic which is computed based on the number of time steps T and the values of T^{shift} and T^{ex}. The column "# Subpr." represents the number of subproblems which are solved during the execution. As indicated in the table, the number of iterations and the number of subproblems do not coincide in general, in contrast to the deterministic case. Recall that in each iteration r, the fixation of further variables results in a decomposition of the current problem P_r into several subproblems which can be solved independently, compare Figure 5.3.

The results of Table 8.13 show that for the first five instances the algorithm performances well with respect to the quality of the generated solutions as well as to the running time. In particular, the relative gap to the optimal solution or to the best lower bound is less than 0.2 % for all selected parameter combinations. Additionally, we detect a clear decrease in the running time if the number of subproblems is reduced. In contrast to the weak effect of the parameter variations for the first five instances, we identify a positive effect of smaller values of T^{ex} and T^{shift} with respect to the solutions' quality for the last instances. This observation is caused by the higher number of facilities of the underlying energy system, compare Table 8.1, yielding subproblems of larger size. Taking the results of all instances into account, the setting of $T^{ex} = 24$ and $T^{shift} = 12$ seems a suitable choice aiming at a reliable generation of good quality solutions.

Limiting the Running Time of the Subproblems

A further aspect affecting the performance of the heuristic concerns the restriction of the running time of the single subproblems occurring during the solution process. Motivated by the good results obtained in the deterministic case shown in Table 8.10, we specify a time limit t_{max} for the solution of each subproblem.

One important aspect influencing the size of the subproblem constitutes the size of the underlying scenario tree. Hence, we decide to take the number of

Table 8.13: Computational results of the approximate-and-fix heuristic for determining T^{ex} and T^{shift}

Instance	T^{Shift}	T^{Ex}	# Iter.	# Subpr.	Upper Bd.	Time	Gap %
tuneInst1	6	12	8	17	573321.8	23.5	0.16
	12	24	4	6	573401.7	24.1	0.17
	24	48	2	2	573004.5	35.2	0.10
tuneInst2	6	12	11	17	281166.0	8.9	0.02
	12	24	6	10	281159.7	7.5	0.01
	24	48	3	4	281300.6	11.5	0.06
	36	60	2	3	281247.9	10.8	0.05
tuneInst3	6	12	11	17	169993.0	8.2	0.05
	12	24	6	10	169941.7	7.4	0.02
	24	48	3	4	169995.7	12.6	0.05
	36	60	2	3	169923.1	9.1	0
tuneInst4	6	12	16	76	1784648.0	201.0	0.11
	12	24	8	37	1783972.7	145.9	0.08
	24	48	4	18	1784205.9	131.8	0.09
	48	72	2	5	1783693.8	100.4	0.06
	60	84	2	5	1784067.5	100.2	0.08
tuneInst5	6	12	16	124	626314.3	719.7	0.08
	12	24	8	57	626069.3	478.4	0.04
	24	48	4	23	625916.0	304.4	0.01
	48	72	2	5	625990.8	221.7	0.03
	60	84	2	5	625964.4	204.9	0.02
tuneInst6	6	12	11	21	1887987.4	184.3	0.49
	12	24	6	12	1885443.2	148.5	0.255
	24	48	3	6	1883877.8	148.7	0.27
	36	60	2	2	1938567.7	94.8	3.08
tuneInst7	6	12	11	21	1783874.9	42.2	0.04
	12	24	6	12	1784230.1	38.9	0.06
	24	48	3	6	1784154.9	40.1	0.05
	36	60	2	2	1784478.0	86.2	0.07
tuneInst8	6	12	11	21	498657.9	247.8	0.73
	12	24	6	12	499677.9	175.2	0.93
	24	48	3	6	504139.5	106.3	1.80
	36	60	2	2	504588.2	152.7	1.89

Table 8.14: Computational results of the approximate-and-fix heuristic for determining the scaling factor k

Instance	k	# Iter.	# Subpr.	Upper Bd.	Time	Gap %
tuneInst1	10	4	6	572528.8	23.7	0.02
	50	4	6	573432.8	24.1	0.18
	100	4	6	574435.0	23.7	0.255
tuneInst2	10	6	10	281159.7	8.6	0.01
	50	6	10	281159.7	8.4	0.01
	100	6	10	282334.3	7.8	0.43
tuneInst3	10	6	10	169941.7	8.0	0.02
	50	6	10	169941.7	8.1	0.02
	100	6	10	169941.7	8.0	0.02
tuneInst4	10	8	37	1784056.3	169.7	0.08
	50	8	37	1784119.6	165.4	0.08
	100	8	37	1783972.7	153.3	0.08
tuneInst5	10	8	57	626041.6	554.1	0.03
	50	8	57	626097.0	560.7	0.04
	100	8	57	626100.8	571.3	0.04
tuneInst6	10	6	12	1887918.3	361.5	0.48
	50	6	12	1883802.7	317.3	0.26
	100	6	12	2383348.9	157.9	21.17
tuneInst7	10	6	12	1783996.1	89.7	0.05
	50	6	12	1784218.1	43.2	0.06
	100	6	12	1784714.9	43.7	0.09
tuneInst8	10	6	12	498338.0	406.4	0.66
	50	6	12	536498.2	177.3	7.73
	100	6	12	2211035.7	126.3	77.61

nodes of the tree into consideration, which we denote by N. Furthermore, the number of facilities which are considered in the energy system plays an important role. We remark that the problem size significantly increases with an augmented number of facilities which are represented by the parameters n_C, n_G, n_P, and n_A. Finally, we account for the total number R of subproblems occurring during the execution, which also gives an indication for the problem size. Altogether and in analogy to the formula (8.2), we set the time limit parameter to

$$t_{max} = \frac{N(n_C + n_G + n_P + n_A)}{k \log(R)}. \tag{8.2}$$

The number R of subproblems is included logarithmically since this means that the resulting time limit is still sufficiently large if the number of subproblems increases significantly assuming a constant number of nodes. Indeed, having chosen T^{ex} and T^{shift}, a high number of subproblems for a fixed number of nodes is an indication for a dense scenario tree which is computational challenging. The parameter $k \in \mathbb{R}_+$ represents a scaling factor allowing to adapt this computation to the problem under consideration. Hence, the value of k is chosen based on the same set of instances used in the previous table.

For the computations we consider the values $k = 10, k = 50$, and $k = 100$, applying the default parameter setting described above. Table 8.14 lists the computational results following the denotation of the previous table. Comparing the relative gap of the solutions generated by the heuristic, we observe a clear trend of decreasing quality with increasing k, although the running time is shortened as intended by the use of the scaling factor. The selection of an appropriate default value for k is mainly based on the results of instances tuneInst6 and tuneInst8 showing the strongest change in quality. Consequently, we choose $k = 50$ for the following computations.

Applying the Approximate-and-Fix Heuristic to Large Instances

Having determined a general setting for the parameters guiding the approximate-and-fix heuristic, in the following we investigate its applicability for the solution of large instances. In summary, we apply the following methods and parameter values:

- We set $T^{shift} = 12$ and $T^{ex} = 24$.
- We apply the approximation strategy $\tilde{S}_1$.
- We set the time scaling factor for each subproblem to $k = 50$.

On this basis, we solve ten large problem instances comparing the results with the exact solver CPLEX which are summarized in Table 8.15. The computations are performed in analogy to the test runs of the rolling horizon heuristic shown in Table 8.11. As for the deterministic case, we assume a

Table 8.15: Computational results of the approximate-and-fix heuristic applied to large instances

Instance	System	# Var.	# Bin.	# Con.	Meth.	Upper Bd.	Time	Gap %
instS1_96	(1111)	12296	5380	11337	*A&F*	450853.0	21.9	0
					Ex	450842.2	693.9	0
instS2_144	(1111)	18440	8068	17001	*A&F*	721023.5	29.4	0.06
					Ex	720571.2	10000.0	0.01
instS3_216	(1111)	54792	23972	50525	*A&F*	1300903.1	198.2	0.10
					Ex	1299478.0	10000.0	0.02
instS4_288	(1111)	135176	59140	124633	*A&F*	1661523.1	1166.7	0.15
					Ex	1659923.4	10000.0	0.06
instS5_60	(1133)	38626	16856	35375	*A&F*	867112.6	98.5	0.91
					Ex	863852.1	10000.0	0.54
instS6_60	(3311)	21074	9368	19675	*A&F*	2929812.0	15.9	0
					Ex	2929740.1	242.1	0
instS7_60	(3333)	44716	19668	41229	*A&F*	4242448.9	100.5	0.03
					Ex	4242340.25	10000.0	0.03
instS8_60	(5555)	74448	32780	68637	*A&F*	7611657.8	202.0	0.05
					Ex	7612740.4	10000.0	0.07

time limit of 10000 CPU seconds and consider instances which are scaled with respect to the number of time steps as well as to the number of facilities.

Considering the quality of the solutions generated by the approximate-and-fix heuristic, we observe that for all instances nearly optimal solutions have been found. Except for one case, the relative gap is less than 0.15 %, which confirms the proposed suitability of this approach for the application to the *S-OPGen* problem. Nevertheless, the running times of the heuristic show a strong increase if the number of time steps is augmented. This observation is based on the significant increase of the number of nodes of the underlying scenario tree, which is reflected by the strongly growing number of variables and constraints shown in column four and six. However, in case larger planning horizons are considered, the incorporation of a fixed approximated period may become reasonable as in the deterministic case. Regarding the last three instances, the heuristic performs remarkably well with respect to the quality and the running time although the number of facilities has been increased. Furthermore, we point out the results of instance instStoch8, where the approximate-and-fix heuristic determines a feasible solution of lower objective function value than the one obtained by CPLEX within the time limit of 10000 seconds. Though, the running times of CPLEX and the approximate-and-fix heuristic are not comparable, since the former constitutes an exact solver which aims at giving quality certificates rather than only focusing on the generation of good quality solutions. Nevertheless, the approximate-and-fix heuristic shows a considerably good overall performance, being able to determine almost optimal solutions in acceptable running time. These results provide the basis for the decision to incorporate this approach into the *SD-BB* framework as initial construction heuristic.

8.4 *SD-BB* Algorithm

We now turn our attention to the main part of this chapter, where we investigate the computational behavior of the *SD-BB* algorithm introduced in Chapter 6. To this end, a series of test runs is performed applying our implementation described in Chapter 7. First, we aim at deriving general suggestions for an adequate parameter setting and branching rule selection. Subsequently, the algorithm is applied to large instances in order to evaluate the performance of the algorithm based on the chosen parameter combination.

For the calibration we use the following parameters and methods as standard setting. The original subproblem is reformulated based on the subdivision of the corresponding scenario tree, whose implementation is explained in Section 7.1. We start the parameter tuning with the determination of a suitable number of subtrees in Section 8.4.1. Furthermore, we determine a feasible solution by the approximate-and-fix heuristic with the default parameter setting obtained in the previous section. For the generation of feasible solutions during the optimization process, the construction heuristic described in Section 7.4.2 is executed exploiting local information of the current branch-and-bound node. With respect to the minimum runtime and downtime restrictions, we follow the suggestion of Section 8.2 of omitting the original constraints from the model and separating them during the optimization process of the subproblems. In case of the branch-and-bound framework, we ensure these constraints implicitly in the incumbent callback, rejecting candidates for a feasible solution in case of a violation. Since the separation would result in additional time-connecting constraints, we have decided to choose this approach. Another motivation for this decision is given by the integer feasibility of the solutions computed for the lower bound and the incorporation of start-up costs. A suitable frequency of the call of the heuristic is determined in Section 8.4.3. Additionally, we execute the *SD-BB* approach extended by Lagrangian relaxation as introduced in Section 6.4. The resulting improvement of the lower bound compared to the complete relaxation together with the results of the corresponding subgradient method are summarized in Section 8.4.2. As default branching strategy, we apply the maximal violation approach which is compared to the strong branching version in the Section 8.4.4.

The test runs for the calibration of the algorithm are performed based on a set of eight instances, which are described in detail in Section 8.1.3. In order to assess the quality and performance of the *SD-BB* algorithm, the results are compared with the solution obtained by applying CPLEX to the mixed-integer program. The results for the tuning instances obtained by CPLEX can be found in Table 8.7. For all running times concerning the calibration of the *SD-BB* algorithm, we set a time limit of 3600 CPU seconds.

8.4.1 Decomposing the Scenario Tree

First, we concentrate on the determination of an appropriate number K of subtrees into which the scenario tree is divided. Recall that based on this subdivision, the original problem is decomposed into K independent subproblems, as described in Section 6.2. By restoring the corresponding coupling constraints, the subproblems are reconnected yielding a reformulation of the original problem. Hence, the choice of K influences the size of the resulting subproblems on the one hand and affects the number of coupling constraints on the other hand, while both aspects have a strong impact on the performance of the *SD-BB* algorithm.

The computational results of the *SD-BB* algorithm for different values of K are shown in Tables 8.16 and 8.17. Next to the number of nodes of the scenario tree shown in the second column, the varying number of subtrees is listed in the third one. In detail, we consider values ranging from 2 through 18 depending on the size of the instance under consideration. The fourth column comprises the number of branch-and-bound nodes created by the *SD-BB* algorithm during the execution. The best lower bound computed on the basis of the decomposed problem is shown in column "Lower Bd.", while the best feasible solution found is presented in column "Upper Bd.".

Table 8.16: Computational results for determining the number of subtrees for tuneInst1 to tuneInst5

Instance	N	#Subtr.	#Nodes	Lower Bd.	Upper Bd.	Time	Gap%
tuneInst1	144	2	10	572239.5	572287.4	559.6	0
		3	15	572217.4	572217.4	365.0	0
		4	20	572106.7	572106.7	203.6	0
		6	75	571893.2	571893.2	328.3	0
		8	564	571949.8	572005.9	1197.3	0
		10	383	571848.9	571855.5	606.6	0
		12	559	571552.1	571552.1	873.6	0
tuneInst2	90	2	152	281070.8	281070.8	746.9	0
		3	1065	280772.6	281329.4	3600.0	0.20
		4	4103	280031.8	281091.5	3600.0	0.258
		6	6572	279943.1	281159.7	3600.0	0.43
		8	6957	279753.9	281159.7	3600.0	0.50
		10	7841	279785.0	280461.7	3600.0	0.24
		12	8451	279806.8	281159.7	3600.0	0.49
tuneInst3	90	2	148	169903.7	169903.7	537.8	0
		3	691	169790.25	169884.2	3600.0	0.06
		4	2976	169454.9	169879.9	3600.0	0.25
		6	6320	169014.3	169905.2	3600.0	0.53
		8	7486	168826.8	170059.6	3600.0	0.72
		10	7790	166529.7	170059.6	3600.0	2.12
		12	8188	168913.0	170059.6	3600.0	0.67
tuneInst4	371	2	37	1736817.7	1783972.7	3600.0	2.72
		3	58	1716312.4	1783972.7	3600.0	3.79
		4	30	1783024.1	1783143.6	1326.7	0
		6	156	1780995.4	1783311.2	3600.0	0.13
		8	309	1782888.4	1783086.6	3600.0	0.01
		10	406	1782548.2	1782953.8	3600.0	0.02
		12	526	1782121.2	1783042.8	3600.0	0.05
tuneInst5	823	2	74	625578.1	625898.9	3600.0	0.05
		4	54	625524.8	625861.5	3600.0	0.06
		8	96	625656.7	625803.4	3600.0	0.02
		10	90	625547.4	625834.3	3600.0	0.05
		12	109	625643.6	626001.6	3600.0	0.06
		15	140	625939.5	625639.5	3600.0	0.06
		18	187	625619.8	625835.5	3600.0	0.03

Table 8.17: Computational results for determining the number of subtrees for tuneInst6 to tuneInst8

Instance	N	#Subtr.	#Nodes	Lower Bd.	Upper Bd.	Time	Gap %
tuneInst6	90	2	22	1542081.4	2343577.6	3600.0	51.97
		3	30	1878249.5	1884991.6	3600.0	0.256
		4	24	1878260.1	1884991.6	3600.0	0.256
		6	38	1877709.9	1884185.5	3600.0	0.254
		8	137	1862572.5	2354623.3	3600.0	26.42
		10	76	1876166.6	2354623.3	3600.0	25.50
		12	126	1869322.3	2354125.8	3600.0	25.93
tuneInst7	90	2	18	1783043.3	1783760.51	3600.0	0.04
		3	78	1783170.6	1783699.4	3600.0	0.03
		4	145	1783259.9	1783632.1	3600.0	0.02
		6	404	1783190.1	1783512.1	3600.0	0.02
		8	550	1783069.1	1783377.5	3600.0	0.02
		10	680	1783011.2	1783253.8	3600.0	0.01
		12	839	1782801.7	1783756.1	3600.0	0.05
tuneInst8	90	2	10	490144.5	499684.9	3600.0	1.95
		3	17	493279.0	499684.9	3600.0	1.30
		4	19	493114.3	499677.9	3600.0	1.33
		6	23	493449.2	499684.9	3600.0	1.26
		8	32	493129.6	499684.9	3600.0	1.33
		10	56	493226.7	499684.9	3600.0	1.31
		12	59	493214.7	499684.9	3600.0	1.31

The last two columns document the CPU seconds spent for the solution and the relative gap resulting from the best lower and upper bound.

Regarding the results of the first five tuning instances, the selection of the number of subproblems shows only a minor impact on the relative gap between the lower and upper bound. With the exception of three instances among those varying in the number of subtrees, all selected values yield a gap of less than 0.8 % within the running time limit of one hour. Nevertheless, for instances tuneInst1, tuneInst2, and tuneInst3 exactly one value leads to the optimal solution. Considering instance tuneInst4, we observe a clear trend suggesting the selection of a larger number of subtrees than for the first three tuning instances. Taking instances with a higher number of energy storages into account, i.e., tuneInst6 and tuneInst8, the relative gap clearly increases. Recall that a higher number of energy storages yields an augmented number of continuous splitting variables posing a further challenge for the *SD-BB* algorithm. Although the results do not show a significant preference for the number of subtrees, they indicate a dependence on the number of nodes of the scenario tree. For the first five instances including only four facilities in the corresponding generation system, we observe that subproblems relying on subtrees of approximately $p = 40$ nodes yield satisfying results. Regarding instances of more facilities, this number should be reduced to $p = 20$ for a good performance. On this basis we choose to set the number of subtrees to $K = \left\lceil \frac{N}{p} \right\rceil$.

Once we have determined the number of subtrees K, the *SD-BB* computes a first lower bound on the optimal objective function value, which is computationally investigated in the following section.

8.4.2 Computing a First Lower Bound

As described in Section 6.4, we extend the basic *SD-BB* approach by the application of Lagrangian relaxation aiming at generating tight lower bounds of the problem at an early stage of the solution process. Recall that this approach has an impact on the formulation of the according subproblems and hence is applied only in the root node of the *SD-BB* algorithm. In order to increment the lower bound, we are interested in determining good values for the corresponding Lagrangian multipliers. Therefore, we make use of the commonly applied subgradient method whose implementation is explained in Section 7.3.1. In the following, we investigate the performance of the subgradient method in order to select a suitable iteration limit $\bar{R}$. Since we make use of this method within the initialization of the *SD-BB* algorithm, the goal is to generate a tight lower bound with low computational effort, i.e., the iteration limit and the quality of the lower bound have to be balanced carefully.

To this end, we apply the subgradient method to the eight tuning instances used in the previous section. We choose an initial step length of $\mu_0 = 0.1$ and set $N = 2$ for the updating step defined in formula (7.8). The results for the first 50 iterations are shown in Table 8.18 and Table 8.19, where they are compared with the results of the basic version which completely neglects the coupling constraints. The method under consideration is described in the second column where "Complete Relax." refers to the complete neglect of the coupling constraints and "Lagrangian Relax." to the Lagrangian relaxation, respectively. In case of the application of the latter method, the

Table 8.18: Computational results for determining an iteration limit of the subgradient method for tuneInst1 to tuneInst5

Instance	Method	# Iter.	Lower Bd.	Time	Gap %
tuneInst1	Complete Relax.	-	443416.2	99.55	22.68
tuneInst1	Lagrangian Relax.	0	571925.5	12.4	0.27
	Lagrangian Relax.	1	572005.1	29.9	0.25
	Lagrangian Relax.	5	572128.2	110.1	0.23
	Lagrangian Relax.	10	572160.7	204.7	0.23
	Lagrangian Relax.	20	572179.8	414.7	0.22
	Lagrangian Relax.	50	572195.9	1094.9	0.22
tuneInst2	Complete Relax.	-	249487.9	6.5	11.26
tuneInst2	Lagrangian Relax.	0	280598.5	6.5	0.19
	Lagrangian Relax.	1	280675.1	10.4	0.16
	Lagrangian Relax.	5	280683.5	25.9	0.16
	Lagrangian Relax.	10	280690.25	45.2	0.16
	Lagrangian Relax.	20	280662.2	84.0	0.17
	Lagrangian Relax.	50	280685.6	199.2	0.16
tuneInst3	Complete Relax.	-	139827.5	4.7	17.72
tuneInst3	Lagrangian Relax.	0	169675.0	3.2	0.16
	Lagrangian Relax.	1	169775.2	6.5	0.10
	Lagrangian Relax.	5	169776.2	17.6	0.10
	Lagrangian Relax.	10	169777.7	29.2	0.10
	Lagrangian Relax.	20	169785.5	53.7	0.09
	Lagrangian Relax.	50	169786.2	128.4	0.09
tuneInst4	Complete Relax.	-	1524676.7	66.51	14.53
tuneInst4	Lagrangian Relax.	0	1780900.25	40.4	0.12
	Lagrangian Relax.	1	1780644.3	64.9	0.13
	Lagrangian Relax.	5	1780754 .9	186.1	0.13
	Lagrangian Relax.	10	1780910.4	318.1	0.12
	Lagrangian Relax.	20	1780952.2	586.0	0.12
	Lagrangian Relax.	50	1780928.7	1391.1	0.12
tuneInst5	Complete Relax.	-	553247.3	78.2	11.99
tuneInst5	Lagrangian Relax.	0	625087.6	20.4	0.10
	Lagrangian Relax.	1	625084.2	38.9	0.10
	Lagrangian Relax.	5	625117.9	107.2	0.10
	Lagrangian Relax.	10	625131.5	190.4	0.10
	Lagrangian Relax.	20	625145.6	367.1	0.09
	Lagrangian Relax.	50	625153.4	914.0	0.09

Table 8.19: Computational results for determining an iteration limit of the subgradient method for tuneInst6 to tuneInst8

Instance	Method	# Iter.	Lower Bd.	Time	Gap %
tuneInst6	Complete Relax.	-	711614.0	101.6	62.24
tuneInst6	Lagrangian Relax.	0	1877348.3	80.4	0.258
	Lagrangian Relax.	1	1842243.2	132.0	2.24
	Lagrangian Relax.	5	1867250.25	392.6	0.92
	Lagrangian Relax.	10	1870195.4	703.4	0.76
	Lagrangian Relax.	20	1877074.9	1322.6	0.40
	Lagrangian Relax.	50	1877620.5	3182.3	0.257
tuneInst7	Complete Relax.	-	1632426.1	24.0	8.51
tuneInst7	Lagrangian Relax.	0	1780718.2	8.2	0.14
	Lagrangian Relax.	1	1780787.2	19.7	0.14
	Lagrangian Relax.	5	1781048.5	47.2	0.13
	Lagrangian Relax.	10	1781116.3	83.1	0.12
	Lagrangian Relax.	20	1781123.7	152.9	0.12
	Lagrangian Relax.	50	1781132.7	361.0	0.12
tuneInst8	Complete Relax.	-	385446	101.9	22.86
tuneInst8	Lagrangian Relax.	0	493355.4	104.2	1.27
	Lagrangian Relax.	1	492161.1	206.0	1.51
	Lagrangian Relax.	5	492894.8	611.6	1.36
	Lagrangian Relax.	10	492806.1	1119.6	1.38
	Lagrangian Relax.	20	492871.3	2130.25	1.36
	Lagrangian Relax.	50	492887.7	5160.4	1.36

third column shows the current iteration of the subgradient method. Recall that for both methods mixed-integer subproblems have to be solved. In particular for the subgradient method, they are solved once in each iteration. The current lower bound is shown in the fourth column and the column before last contains the CPU time consumed for the computation. Using the upper bound determined by the approximate-and-fix heuristic in the root node of the branch-and-bound tree, the relative gap is computed shown in the last column.

Comparing the lower bound based on the complete relaxation with the lower bound generated in the first iteration of the subgradient method, we observe a strong increase by applying the latter method. This remarkably good result is owed to the suitable starting point of the subgradient method.

Recall that the initial Lagrangian multipliers are chosen based on the dual values of the LP relaxation. Indeed, the subgradient method only slightly improves the lower bound within the next 50 iterations. Furthermore, the consideration of the running time illustrates the high computational effort, since in each iteration all subproblems resulting from the decomposition have to be solved. Consequently, we decide to choose a small value for limiting the iteration number of the subgradient method, i.e., we set $\bar{R} = 5$ for the further computations.

8.4.3 Heuristics

Having discussed the parameter setting for the determination of the lower bound, in the following section we focus on the generation of feasible solutions. Besides the application of the approximate-and-fix heuristic whose calibration has been specified in Section 8.3.2, we include a further heuristic method in the *SD-BB* framework which is responsible for the computation of feasible solutions during the solution process. As described in Section 7.4.2, this heuristic relies on the exploitation of the local information available in the corresponding branch-and-bound node by fixing as much binary variables as possible to values obtained in the current node.

For its incorporation into the *SD-BB* algorithm, we need to choose a suitable frequency defining how often the heuristic is executed during the solution process. Since the execution of the heuristic requires the solution of a mixed-integer program, the frequency has a considerable impact on the performance of the entire algorithm.

For the determination of a suitable frequency, we perform a series of test runs based on the eight tuning instances introduced in Section 8.1.3. In detail, we set the frequencies to 1 and 1/10, which means that the heuristic is used in every branch-and-bound node and in every tenth node, respectively. Additionally, we consider a variant of combining its execution with the branching process. More precisely, we decide to execute the heuristic every time the branching on a continuous variable requires the creation of an additional grid point, as explained in Section 7.2. This approach is motivated by the observation that the refinement increases the possibility of finding a further feasible solution based on the values in the current branch-and-bound node. Furthermore, we choose the distance parameter $d = 3$ as default value being large enough to allow the determination of a feasible solution in most of the cases. Recall that d defines the surrounding of the split nodes in the decomposed scenario tree whose variables are not fixed for

Table 8.20: Computational results of the *SD-BB* algorithm for determining the frequency of the heuristic

Instance	Freq.	# Nodes	Lower Bd.	Upper Bd.	Time	Gap %
tuneInst1	1	79	572106.6	572280.25	3600.0	0.03
	1/10	42	572169.0	572180.2	1644.6	0
	refine	30	572179.2	572189.2	1184.1	0
tuneInst2	1	41	281071.4	281073.4	1674.3	0
	1/10	15	281070.8	281073.8	273.7	0
	refine	15	281070.8	281073.8	268.2	0
tuneInst3	1	25	169901.7	169909.7	612.0	0
	1/10	15	169902.6	169910.4	98.2	0
	refine	15	169903.7	169903.7	98.2	0
tuneInst4	1	55	1782202.4	1783117.7	3600.0	0.05
	1/10	8	1782981.7	1783094.4	905.5	0
	refine	11	1782990.6	1783151.1	509.3	0
tuneInst5	1	54	625611.3	626015.1	3600.0	0.06
	1/10	65	625611.3	626019.9	3600.0	0.06
	refine	67	625612.7	626015.1	3600.0	0.06
tuneInst6	1	25	1877747.7	1883070.0	3600.0	0.28
	1/10	38	1877850.6	1883470.7	3600.0	0.250
	refine	36	1877794.7	1883070.0	3600.0	0.28
tuneInst7	1	144	1783176.1	1783944.4	3600.0	0.04
	1/10	190	1783346.1	1783852.8	3600.0	0.03
	refine	199	1783359.6	1783627.9	3600.0	0.02
tuneInst8	1	21	493747.3	498792.6	3600.0	1.02
	1/10	26	493833.8	498854.8	3600.0	1.02
	refine	30	493862.3	498564.8	3600.0	0.95

the generation of a feasible solution. The computational results are summarized in Table 8.20, where a further column denoted by "Freq." represents the chosen frequency.

Comparing the running time of the first four instances, we detect a significant increase by applying the heuristic in every branch-and-bound node, as expected. The additional executions of the heuristic slow down the solution process without improving the upper bound significantly. Regarding the larger instances, this observation is confirmed by three instances yielding a worse gap for a higher frequency. However, the third strategy based on the refinement during the branching on continuous variables performs best, which is particularly apparent for the first and the last instance. This behavior may result from the coupling of the execution of the heuristic to the local information of the current branch-and-bound node rather than applying an independent frequency strategy. Concluding these results, we decide to set the latter strategy as default method in the *SD-BB* algorithm.

8.4.4 Branching

In this section, we investigate the choice of an appropriate branching rule applied within the *SD-BB* framework. More precisely, we compare the two basic concepts for the variable selections which are explained in Section 7.2.1. Recall that one rule selects a pair of variables which produces the maximum violation of the corresponding coupling constraint, while the second rule relies on the idea of strong branching by performing a one-step look-ahead. Furthermore, we consider the variant of combining both methods by alternating their execution. This combination aims at exploiting the advantages of both approaches which consist of a fast selection of a suitable pair of variables on the one hand and the reduction of the number of branch-and-bound nodes on the other hand.

For the test runs, we make use of the eight tuning instances described above. The results are listed in Table 8.21 where the second column indicates the chosen branching rule. In case of the strong branching approach, we choose a weighting parameter $\mu = \frac{1}{6}$ as proposed by [AKM05].

As expected, the strong branching approach performs well with respect to the number of branch-and-bound nodes evaluated during the solution process. In all cases except for tuneInst6, a significant decrease of the number of nodes can be detected. Nevertheless, for the first instance the maximal violation approach yields an optimal solution in a shorter running time. This effect may be explained by the additional computational effort when applying the strong branching method. Recall that in each branch-and-bound node, for all continuous splitting variables an LP is solved estimating the variation of the objective function value. For the first instance, this property predominates the advantage of evaluating less branch-and-bound nodes. However, this is not the case for the other instances. The pure strong branching approach also shows a better performance than the com-

Table 8.21: Computational results of the *SD-BB* algorithm comparing different branching strategies

Instance	Branching	#Nodes	Lower Bd.	Upper Bd.	Time	Gap%
tuneInst1	max. viol.	45	572106.8	572106.8	259.2	0
	strong	26	572198.2	572198.2	880.8	0
	combined	28	572147.8	572147.8	658.1	0
tuneInst2	max. viol.	323	281070.8	281070.8	1776.4	0
	strong	24	281070.8	281070.8	132.2	0
	combined	25	281070.8	281070.8	137.6	0
tuneInst3	max. viol	161	169903.7	169903.7	531.9	0
	strong	38	169903.7	169903.7	107.2	0
	combined	39	169903.7	169903.7	112.6	0
tuneInst4	max. viol.	209	1782933.9	1783132.3	3600.0	0.01
	strong	16	1782967.1	1783074.1	972.2	0
	combined	33	1783242.9	1783242.9	2119.9	0
tuneInst5	max. viol.	257	625623.0	626041.6	3600.0	0.07
	strong	169	625944.0	626590.5	3600.0	0.10
	combined	187	625619.8	625835.5	3600.0	0.03
tuneInst6	max. viol.	65	1877719.2	1884881.9	3600.0	0.258
	strong	38	1877709.9	1884185.5	3600.0	0.254
	combined	33	1877712.2	1884484.9	3600.0	0.256
tuneInst7	max	346	1783186.6	1783513.5	3600.0	0.02
	strong	49	1783274.9	1783314.5	1134.4	0
	combined	150	1783142.4	1783531.0	3600.0	0.02
tuneInst8	max. viol.	56	493586.7	499685.0	3600.0	1.22
	strong	24	493989.4	498963.7	3600.0	1.00
	combined	47	493533.6	498977.4	3600.0	1.09

bination of both methods except for tuneInst5. The success of the former method may be explained by the expensive computation of a lower bound in each branch-and-bound node. Since the evaluation of a node requires the solution of a mixed-inter program, the reduction of the overall number of branch-and-bound nodes mostly results in a reduction of the entire running time. Motivated by these results, we choose the strong branching rule as default method in the *SD-BB* algorithm.

8.4.5 Accuracy

Having addressed the calibration of the *SD-BB* algorithm, it remains to investigate the effect of imposing different accuracies $\delta > 0$ to the algorithm. Recall that this accuracy concerns the continuous splitting variables which are defined during the reformulation of the original problem. As specified in Section 6.3, δ describes the maximal violation of the coupling constraints allowed during the solution process, i.e., we say that a coupling condition of a continuous pair of variables $(x_n, x_{\tilde{n}})$ is satisfied if $|x_n - x_{\tilde{n}}| \leq \delta$, where n and $\tilde{n}$ denote a split node and the duplicated node, respectively.

Note that by applying this accuracy criterion, we deal with the absolute violation of the constraints. Consequently, this accuracy needs to be chosen in dependence of the values assumed by the affected variables. For the *S-OPGen* problem, the continuous splitting variables simply consist of the variables describing the energy storage levels. Within our applications, the values of these variables vary between 60 and 600. Taking these assumptions into account, we decide to consider three different accuracy levels allowing an absolute violation of $\delta = 1.0$, $\delta = 0.5$, and $\delta = 0.1$. Based on the lower bound of the variables given above, these accuracy levels yield a maximum relative error of approximately $\delta_{rel} = 0.0167, \delta_{rel} = 0.0083$, and $\delta_{rel} = 0.0017$, respectively.

In order to investigate the effect of the accuracies on the solution process of the *S-OPGen* problem, we perform a series of test runs comparing the three different accuracies. Moreover, we take the best upper and lower bound of the original problem into account, which are obtained using CPLEX. The outcome of the solution processes is summarized in Table 8.22.

As expected, the running times of the computations applying a coarser accuracy are shorter than for a finer one. Clearly, restoring the relaxed coupling constraints is facilitated under a higher value of δ. Comparing the objective function values for the chosen accuracies, we detect a slight increase in the lower and upper bound when the accuracy is refined. This effect is reasonable since a higher value of δ results in a further relaxation of the coupling constraints. This observation is emphasized by taking the exact solution into account. For all instances solved to optimality by the *SD-BB* algorithm, the exact best upper bound is slightly higher than the optimal function value obtained using *SD-BB*.

For an estimate of the accuracy error occurring in the objective function value, we compare the solutions of the instances solved to optimality with the upper bound obtained by applying CPLEX. In the worst case, accuracy

Table 8.22: Computational results of the *SD-BB* algorithm comparing different accuracy levels

Instance	Accuracy	Lower Bd.	Upper Bd.	Time	Gap %
tuneInst1	exact	572153.4	572418.3	3600.0	0.05
	$\delta = 0.1$	572343.0	572380.0	1107.5	0
	$\delta = 0.5$	572144.7	572144.7	725.1	0
	$\delta = 1.0$	571844.8	571844.8	593.4	0
tuneInst2	exact	281118.3	281159.7	3600.0	0.01
	$\delta = 0.1$	281142.0	281149.1	542.2	0
	$\delta = 0.5$	281070.8	281075.1	148.4	0
	$\delta = 1.0$	280971.5	280995.6	67.9	0
tuneInst3	exact	169906.4	169923.1	112.3	0
	$\delta = 0.1$	169919.2	169919.2	244.9	0
	$\delta = 0.5$	169903.7	169903.7	123.0	0
	$\delta = 1.0$	169884.3	169884.3	98.2	0
tuneInst4	exact	1782634.5	1783512.4	3600.0	0.05
	$\delta = 0.1$	1782020.5	1783465.7	3600.0	0.03
	$\delta = 0.5$	1782950.2	1783073.7	1176.7	0
	$\delta = 0.1$	1782670.5	1782670.5	1077.0	0
tuneInst5	exact	625662.5	625823.0	3600.0	0.03
	$\delta = 0.1$	625649.8	625857.1	3600.0	0.03
	$\delta = 0.5$	625619.5	625835.1	3600.0	0.03
	$\delta = 1.0$	625558.2	625785.5	3600.0	0.03
tuneInst6	exact	1878825.5	1882847.5	3600.0	0.21
	$\delta = 0.1$	1877672.9	1883144.3	3600.0	0.29
	$\delta = 0.5$	1877490.7	1883029.3	3600.0	0.29
	$\delta = 1.0$	1877694.6	1883029.3	3600.0	0.28
tuneInst7	exact	1783178.6	1783750.2	3600.0	0.03
	$\delta = 0.1$	1783653.1	1783582.5	3600.0	0.06
	$\delta = 0.5$	1783142.4	1783531.0	3600.0	0.02
	$\delta = 1.0$	1783111.7	1783178.2	778.0	0
tuneInst8	exact	495041.9	498016.1	3600.0	0.59
	$\delta = 0.1$	493986.4	498963.7	3600.0	1.00
	$\delta = 0.5$	493635.8	498789.7	3600.0	1.03
	$\delta = 1.0$	493598.6	498552.9	3600.0	0.99

$\delta = 1.0$ yields a relative difference of 0.1 %, accuracy $\delta = 0.5$ lead to a difference of 0.04 %, and $\delta = 0.01$ results a difference of 0.007 %, all of them appearing for tuneInst1. However, we remark that these values only provide an indication of the approximation error, being computed exemplarily for these tuning instances. As the number of relaxed coupling constraints depends on the number of subtrees, this error most likely increases for a larger number of subproblems. Nevertheless, these results are satisfactory advising a δ of this magnitude.

Altogether, we deal with the trade-off between a high accuracy on the one hand and a fast running time on the other hand. Taking both aspects into account, we decide to choose an accuracy of $\delta = 0.5$ for further computations. This choice is further motivated by the observation that this value allows the determination of an optimal solution in four of eight cases in a relatively short running time.

8.4.6 Solving Large Instances

Based on the parameter setting for the *SD-BB* algorithm determined in the previous sections, in the following we focus on the application of the algorithm to larger problem instances. Aiming at a diversified evaluation of this approach, we scale the chosen instances with respect to the following main characteristics which define an instance of the *S-OPGen* problem: At first, we consider a variation in the planning horizon involving an enlargement of the underlying scenario tree. Secondly, we aim at investigating the effect of changing the input data, i.e., considering variations in the load profiles, in the amount of wind power provided as well as in prices for electricity. Finally, we scale the instances with respect to the number of facilities of the considered energy system. We conclude this section by comparing the results obtained by the *SD-BB* algorithm to those determined by the solver CPLEX, providing the possibility of evaluating the performance of the former approach.

For the test runs of this section, we apply the parameter combinations determined above, which are summarized in the following:

- For the computation of the number of subtrees, we choose $p = 40$ for the small energy systems and $p = 20$ for larger ones which contain more than four facilities.
- For determining an initial solution, we apply the approximate-and-fix heuristic.

Table 8.23: Computational results of the *SD-BB* algorithm scaling the planning horizon

Instance	# Var.	# Bin.	# Con.	Lower Bd.	Upper Bd.	Time	Gap %
instT48	6152	2692	5673	151719.91	151719.91	501.8	0
instT60	7688	3364	7089	215359.3	215359.3	247.3	0
instT72	9224	4036	8505	291515.0	291538.0	1029.7	0
instT96	12296	5380	11337	450752.6	450784.2	341.6	0
instT120	15368	6724	14169	538461.3	538506.1	1425.0	0
instT144	18440	8068	17001	720474.5	720535.8	2149.5	0
instT168	23688	10364	21841	964763.9	965229.0	10000.0	0.05
instT216	54792	23972	50525	1298810.5	1300290.0	10000.0	0.11
instT384	288776	126340	266271	1841029.4	1849406.0	10000.0	0.45

- The frequency of the heuristic within the branch-and-bound process is based on the refinement method.
- We set the iteration limit of the subgradient method to $\bar{R} = 5$.
- We apply the strong branching rule in the branching step.
- As accuracy level we choose $\delta = 0.5$.

Due to the larger size of the instances we increase the running time limit to 10000 CPU seconds for each execution.

At first, we turn our consideration to the scaling of the number of time steps. In detail, the planning horizon varies between 48 and 384 times steps where the latter correspond to four days. For the underlying scenario tree, this variation corresponds to an increase from 48 to 2256 nodes and from one to 28 scenarios. Additionally, we assume the standard generation system to consist of a coal power plant, a gas turbine power plant, a pumped hydro storage and a compressed air energy storage. Solving the according problem instances yields the results shown in Table 8.23. In order to indicate the problem size, columns 2 through 4 show the number of variables, binary variables, and constraints, respectively. As in the previous tables, the last four columns list the best lower and upper bound found during the execution, the consumed running time, and the relative gap.

We observe that up to 144 time steps, the algorithm finds the optimal solution before reaching the time limit of 10000 seconds. In general, the running time increases with an augmented number of time steps, as expected,

Table 8.24: Computational results of the *SD-BB* algorithm scaling input data

Instance	Wind	Load	Prices	Lower Bd.	Upper Bd.	Time	Gap %
instD1	0	0	0	574038.1	574038.1	296.6	0
instD2	+	0	0	4963171.5	4963613.0	359.0	0
instD3	0	+	0	8251729.8	8251729.8	88.6	0
instD4	0	0	−	408442.6	408447.6	3299.4	0
instD5	+	0	−	214789.4	209330.8	10000.0	2.50
instD6	+	+	0	4291079.2	4291499.5	139.8	0
instD7	−	+	0	6010377.4	6010915.9	42.3	0
instD8	+	+	−	1447154.6	1447285.2	1642.0	0

however for instT48 and instT72 longer running times than presumed are necessary. In both cases, this effect is caused by a relatively high value of the first upper bound obtained by the approximate-and-fix heuristic showing a gap of about 0.5 %. Nevertheless, the results obtained for the instances are satisfactory, since even for instances comprising a planning horizon of more than one day an almost optimal solution can be obtained.

For analyzing the behavior of the *SD-BB* algorithm under the variation of the input data, we define a standard instance with a predetermined number of facilities and a given scenario tree. Here, the scenario tree is based on 117 nodes with four scenarios and 60 time steps. In this setting, the wind park and the consumers' demand are well dimensioned with respect to the facilities in order to reflect a reliable energy system. Thereupon, the corresponding data is scaled as indicated in columns 2 through 4 of Table 8.24. To be more precise, "0" reflects the standard value, "+" denotes an augmentation, and "-" a reduction, respectively. Since the problem size coincides for all instances, the columns indicating the number of variables and constraints are omitted.

With respect to the running times of Table 8.24, we observe a significant impact of the varied data on the time spent for the solution. Indeed, the running times range between 43 to 10000 CPU seconds, which constitutes the predefined time limit for the computations. Nevertheless, the problem is solved to optimality in seven of eight cases. Furthermore, we can establish a relation between the variations of the price level for electricity and the performance. It becomes apparent that by assuming a lower level of prices for electricity, the solution time significantly increases, see instances instD4, instD5, and instD8. This effect may be caused by the intensified contribu-

Table 8.25: Computational results of the *SD-BB* algorithm scaling the number of facilities

Instance	# Var.	# Bin.	# Con.	Lower Bd.	Upper Bd.	Time	Gap %
instF1100	3163	1406	3044	913476.9	913517.3	1.3	0
instF1111	14984	6556	3821	875068.0	875068.0	207.1	0
instF1133	38626	16856	35375	858916.1	867112.6	10000.0	0.95
instF3300	9253	4218	8898	4275990.4	4275995.8	0.9	0
instF3311	21074	9368	19675	2929812.0	2929812.0	144.4	0
instF5555	74448	32780	68637	7607469.9	7611657.8	10000.0	0.06
instF7733	56896	25292	52937	8996898.6	8997774.9	3409.0	0
instF3377	92000	40268	84337	3522533.1	3534360.8	10000.0	0.33

tion of the power plants and energy storages to the regulation of fluctuating supply and demand rather than procuring energy from the energy market. In particular, the power plants show a more variable behavior which hedges about the decision of their commitment. Furthermore, the results indicate a better performance if the load is scaled up, compare instances instD3 and instD7, and a worse performance in case the available wind power is augmented, see in particular instance instD2. We believe that this effect may result from the strong fluctuating behavior of the wind power, whose effect is strengthened in case it is scaled up. Concluding these results, the variation of data strongly impacts the performance of the algorithm. However, the *SD-BB* algorithm determines the optimal solutions in acceptable running time for the majority of the instances.

The following investigations concern the scaling of the facilities of the underlying generation system. For the computations, we consider a fixed scenario tree of 118 nodes and four scenarios comprising a planning horizon of 60 time steps. On this basis we alter the underlying energy systems by including two up to 20 facilities. Since the coal power plants and gas turbine power plants rely on the same generic description in the model, we always scale both of them with the same factor. The same holds for the different types of energy storages. The results obtained for the eight instances are shown in Table 8.25, where the combination of facilities is encoded in the instance name which is shown in the first column of the table.

Similar to the previous table, we observe a strong variation of the running time for different instances. More precisely, five of the eight instances are solved to optimality, whereas for the remaining three instances, the computations were interrupted after exceeding the time limit of 10000 seconds. Particularly apparent is the relation between the number of energy storages and the running time spent for the corresponding solution. Regarding instF1100, instF1111, and instF1133, the number of energy storages is raised from zero to six yielding a remarkably strong increase in running time. Besides the increase of the problem size, this effect is also caused by the augmented number of splitting variables occurring in the reformulated problem. Since various branching steps are necessary to restore the relaxed coupling constraints, their increase poses a greater challenge to the *SD-BB* algorithm. Recall that the continuous splitting variables result from the modeling of the energy storage levels. Nevertheless, if the number of power plants is also augmented as in instF7733, the *SD-BB* algorithm can determine the optimal solution before the time limit expires. A possible explanation for this behavior is the increased potential of the plants to balance the fluctuations in wind power without changing their operational level significantly. This results in a less variable behavior of energy storages, which facilitates the restoring of the relaxed coupling constraints. Altogether, these results suggest the suitability of the *SD-BB* algorithm for instances with a relatively large number of power plants in relation to the energy storages. Additionally, for all instances, solutions with a relative gap of less than 1 % are obtained.

Finally, we compare the performance of the *SD-BB* algorithm with the commercial solver CPLEX aiming at evaluating the former method. Since we have proposed the suitability of the *SD-BB* algorithm for specific types of instances, we apply both methods to ten test instances in order to investigate this suggestion based on the results of a further solver. In summary, the results of the previous tables indicate the suitability of *SD-BB* to the application of instances showing a small number of facilities. In case the energy system is enlarged, those with a higher proportion of power plants are more favorable to the performance of the algorithm. The outcome of the solution processes is shown in Table 8.26. The first part of the columns states the size of the corresponding instance. Column 5 indicates whether CPLEX or the *SD-BB* algorithm is applied and finally, the last four columns summarize the corresponding computational results.

We observe that in nine out of twelve instances the *SD-BB* algorithm yields a better performance than CPLEX. To be more precise, in case our algorithm finds the optimal solution before the time limit of 10000 CPU seconds expires, the running time is significantly smaller except for the first instance instT72. Here, our solver consumes 100 seconds more than CPLEX. This

Table 8.26: Computational results comparing CPLEX and *SD-BB*

Instance	# Var.	# Bin.	# Con.	Algor.	Lower Bd.	Upper Bd.	Time	Gap %
instT72	9224	4036	8505	CPLEX	291526.1	291556.3	962.1	0
				SD-BB	291515.0	291538.0	1029.7	0
instT96	12296	5380	11337	CPLEX	450796.4	450842.5	693.9	0
				SD-BB	450752.6	450784.2	341.6	0
instT120	15368	6724	14169	CPLEX	538535.2	538621.1	10000.0	0.02
				SD-BB	538461.3	538506.1	1425.0	0
instT144	18440	8068	17001	CPLEX	720468.5	720571.9	10000.0	0.01
				SD-BB	720474.5	720535.8	2149.5	0
instT168	23688	10364	21841	CPLEX	964999.3	965096.2	2135.7	0
				SD-BB	964763.9	965229.0	10000.0	0.05
instD2	18312	8012	16907	CPLEX	4963332.8	4963830.0	194.0	0
				SD-BB	4963171.5	4963613.0	159.0	0
instD4	18312	8012	16907	CPLEX	407740.5	408381.4	10000.0	0.21
				SD-BB	408442.6	408447.6	3299.4	0
instD6	18312	8012	16907	CPLEX	4291181.8	4291547.0	263.6	0
				SD-BB	4291079.2	4291499.5	139.8	0
instD7	18312	8012	16907	CPLEX	6010647.6	6011157.9	377.1	0
				SD-BB	6010377.4	6010915.9	42.3	0
instF1111	14984	6556	13821	CPLEX	875272.3	875359.1	9963.6	0
				SD-BB	875068.0	875068.0	207.1	0
instF1133	38626	16856	35375	CPLEX	859194.2	863852.2	10000.0	0.54
				SD-BB	858916.1	867112.6	10000.0	0.95
instF3311	21074	9368	19675	CPLEX	2929450.2	2929740.1	242.1	0
				SD-BB	2929812.0	2929812.0	144.4	0

observation results from the relatively high lower bound in the root node of the corresponding scenario tree hedging about closing the gap for this test run. Additionally, the *SD-BB* method is able to determine an optimal solution in ten out of twelve instances while CPLEX solves only eight up to optimality. The superiority of this approach in the running time spent for solution becomes apparent when considering the geometric mean. In detail, the mean running time for the solutions of these instances spent by CPLEX spans 1718.5 CPU seconds, while our *SD-BB* approach shows a mean value of 681.3 CPU seconds, yielding a reduction of more than 50 %. Altogether, the results confirm our claim of suggesting the application of the *SD-BB* approach to these types of problem instances.

Summarizing the computational results of this section, we believe that we have developed a promising approach for solving the *S-OPGen* problem. However, we also detected its difficulties in solving instances in case the number of continuous splitting variables is significantly increased. On the other hand, taking these results into account, the algorithm shows the potential to be successfully applied to a wider range of problems which inherit certain characteristics, as described above.

From the application side, the interest lies on the question of how energy storages may contribute to the decoupling of fluctuating supply and demand. Considering the results shown in Table 8.25, the potential benefit of additional energy storages in an energy generation system is indicated by the decrease of the operational costs. However, for a reliable evaluation, further aspects need to be taken into account as for instance the investment costs as well as the determination of suitable storage dimensions. The energy economical interpretation is carried out by our project partners from the Ruhr-Universität Bochum and the Universität Duisburg-Essen. First results have been published in [EMM+09] analyzing the possibilities of energy storages in the scope of a rising participation of electricity production based on wind power.

Chapter 9

Conclusions

In this thesis, we have developed a novel scenario tree-based decomposition approach incorporated in a branch-and-bound framework for the solution of multistage stochastic mixed-integer programs. This study has been motivated by the real world problem arising in energy production when large amounts of fluctuating energy are fed into the public supply network. Due to the rising participation of electricity based on wind power, the potential of energy storages to decouple fluctuating supply and demand is of great interest.
Within this scope, we have considered a power generation system including conventional power plants, energy storages, and a wind park. The underlying power generation problem is formulated as a mixed-integer optimization program taking the partial load efficiencies as well as the combinatorial aspects of the units into account. The crucial part of the model constitutes the inclusion of uncertainty concerning the amount of available wind power and the prices for electricity purchased on the spot market. By describing the evolution of the uncertain data via a scenario tree, we have formulated a multistage stochastic mixed-integer problem of high complexity.

Under the exploitation of specific structures inherent in this problem, we have developed a decomposition-based solution approach called *SD-BB* relying on the decomposition of the scenario tree into subtrees. On this basis, the problem is reformulated by a set of independent subproblems coupled by few time-connecting constraints. The feasibility of the solutions is recovered by incorporating the approach into a branch-and-bound framework.

In order to make this approach more efficient, the development of several methods has been necessary. To this end, we have constructed a polynomial time algorithm for a fast decomposition of the subtrees yielding a suitable subdivision for our approach. An essential contribution to the good performance of the algorithm constitutes the extension by a Lagrangian relaxation approach. Additionally, we have investigated the polyhedral substructure arising in the course of the minimum runtime and downtime restriction in a scenario tree-based formulation and successfully integrated the obtained facets as cutting planes. For the determination of a feasible solution, an approximate-and-fix heuristic has been designed which has shown remarkably good results for the application to large instances. Furthermore, adapted branching rules have been established focusing on a suitable variable selection and the branching on continuous variables. The algorithm has been implemented in such a way that subproblems which have already been solved earlier in the solution process can be restored in order to avoid redundant solutions of similar subproblems.

We have evaluated the performance of the *SD-BB* approach based on a series of test runs considering instances of different characteristics. Thereupon, we proposed a general setting of parameters and methods for the application to further instances. Concluding these results, the *SD-BB* algorithm is able to solve large instances with up to four days to optimality or at least to provide a quality certificate of a relative gap less than 1%. The results obtained by our approach are also compared to the standard commercial solver CPLEX, indicating the suitability of the *SD-BB* for the solution of these problem instances.

Although we conceived the *SD-BB* algorithm for the solution of the energy production problem described above, its general framework is applicable to a wide range of related problems. However, an adaptation of the implementations becomes necessary, since some of the developed methods applied in the algorithm have been designed specifically for the solution of the *S-OPGen* problem.

Besides the studied methods, this novel approach offers several aspects for further research and improvement. Concerning the performance of the algorithm, the transformation of selected routines applied in a standard branch-and-bound approach such as preprocessing techniques, node selection priorities and more sophisticated branching rules provide a great potential for an additional reduction in running time.

Another point for further research concerns the handling of continuous splitting variables resulting from the reformulation of the problem. Since we have detected an apparent increase in the running time if the number of continuous splitting variables is augmented, a more elaborate handling may enable a successful application to a wider range of problems. In this context, it would be interesting to apply the *SD-BB* algorithm to multistage optimization problems where the time-connecting variables comprise binary variables only. Motivated by the numerical results, we believe that our approach is especially promising for the application to this kind of problems.

From the energy economical point of view, a further direction of research concerns the extension of the model towards the additional dimensioning of the storage sizes taking operational as well as investment costs into account.

Bibliography

[AAEG$^+$03] A. Alonso-Ayuso, L.F. Escudero, A. Garin, M.T. Ortuño, and G. Pérez. An approach for strategic supply chain planning under uncertainty based on stochastic 0-1 programming. *Journal of Global Optimization*, 26(1):97–124, 2003.

[AAEO00] A. Alonso-Ayuso, L.F. Escudero, and M.T. Ortuño. A stochastic 0-1 program based approach for the air traffic flow management problem. *European Journal of Operational Research*, 120:47–62, 2000.

[AAEO03] A. Alonso-Ayuso, L.F. Escudero, and M.T. Ortuño. BFC, A branch-and-fix coordination algorithmic framework for solving some types of stochastic pure and mixed 0-1 programs. *European Journal of Operational Research*, 151:503–519, 2003.

[ABCC07] D.L. Applegate, R.E. Bixby, V. Chvátal, and W.J. Cook. *The Traveling Salesman Problem: A Computational Study (Princeton Series in Applied Mathematics)*. Princeton University Press, 2007.

[AC00] J.M. Arroyo and A.J. Conejo. Optimal response of a thermal unit to an electricity spot market. *IEEE Transactions on Power Systems*, 15(3):1098–1104, 2000.

[AKM05] T. Achterberg, T. Koch, and A. Martin. Branching rules revisited. *Operations Research Letters*, 33:42–54, 2005.

[Bak77] K.R. Baker. An experimental study of rolling schedules in production planning. *Decision Sciences*, 8:19–27, 1977.

[Bal95] R. Baldick. The generalized unit commitment problem. *IEEE Transactions on Power Systems*, 10(1):465–475, 1995.

[BGGG06] P. Beraldi, G. Ghiani, A. Grieco, and E. Guerriero. Fix and relax heuristic for a stochastic lot-sizing problem. *Computational Optimization and Applications*, 33:303–318, 2006.

[BL97] J.R. Birge and F. Louveaux. *Introduction to Stochastic Programming*. Springer Verlag, 1997.

[BLSP83] D.P. Bertsekas, F.S. Lauer, N.L. Sandell, and T.A. Posbergh. Optimal short-term scheduling of large-scale power systems. *IEEE Transactions on Automatic Control*, 28:1–11, 1983.

[BMW] Energie in Deutschland: Trends und Hintergründe zur Energieversorgung in Deutschland. Bundesministerium für Wirtschaft und Technologie (BMWi). `http://www.bmwi.de/navigation/service/publikationen`.

[BSP82] R.I. Becker, S.R. Schach, and Y. Perl. A shifting algorithm for min-max tree partitioning. *Journal of the ACM*, 29:58–67, 1982.

[BT70] E.L.M. Beale and J.A. Tomlin. Special facilities in a general mathematical programming system for nonconvex problems using ordered sets of variables. In *Proceedings of the Fifth International Conference on Operations Research*, pages 447–454, 1970.

[CHS02] S. Chand, V. Hsu, and S. Sethi. Forecast, solution, and rolling horizons in operations management problems. *Manufacturing & Service Operations Management*, 4(1):25–43, 2002.

[CL] T. Christof and A. Löbel. PORTA - POlyhedron Representation Transformation Algorithm. `http://www.zib.de/Optimization/Software/Porta`.

[CPL] ILOG CPLEX Division. Information available at URL `http://www.cplex.com`.

[CS98] C.C. Carøe and R. Schultz. A two-stage stochastic program for unit commitment under uncertainty in a hydro-thermal power system. In *Preprint SC 98-11*, 1998.

[CS99] C.C. Carøe and R. Schultz. Dual decomposition in stochastic integer programming. *Operations Research Letters*, 24:37–45, 1999.

[Dan63] G.B. Dantzig. *Linear Programming and Extensions*. Princeton University Press, 1963.

[Den05] Energiewirtschaftliche Planung für die Netzintegration von Windenergie in Deutschland an Land und Offshore bis zum Jahr 2020. Deutsche Energie-Agentur GmbH (dena), 2005.

[DEWZ94] C. Dillenberger, L.F. Escudero, A. Wollensak, and W. Zhang. On practical resource allocation for production planning and scheduling with period overlapping setups. *European Journal of Operational Research*, 75:275–286, 1994.

[DGKR03] J. Dupacova, N. Gröwe-Kuska, and W. Römisch. Scenario reduction in stochastic programming: An approach using probability metrics. *Mathematical Programming*, 95(3):493–511, 2003.

[EEG00] Gesetz für den Vorrang Erneuerbarer Energien (Erneuerbare-Energien-Gesetz - EEG). BGBl. I, Nr. 13, 2000.

[EKR+07] A. Epe, C. Küchler, W. Römisch, S. Vigerske, H.-J. Wagner, C. Weber, and O. Woll. Stochastische Optimierung mit rekombinierenden Szenariobäumen - Analyse dezentraler Energieversorgung mit Windenergie und Speichern. In *Optimierung in der Energiewirtschaft*, VDI-Berichte 2018, pages 3–13. VDI-Verlag 2007, 2007.

[EKR+09] A. Epe, C. Küchler, W. Römisch, S. Vigerske, H.-J. Wagner, C. Weber, and O. Woll. Optimization of dispersed energy supply - stochastic programming with recombining scenario trees. In J. Kallrath, P.M. Pardalos, S. Rebennack, and M. Scheidt, editors, *Optimization in the Energy Industry*. Springer, 2009.

[EMM+09] A. Epe, D. Mahlke, A. Martin, H.-J. Wagner, C. Weber, O. Woll, and A. Zelmer. Betriebsoptimierung zur ökonomischen Bewertung von Speichern. In R. Schultz and H.-J. Wagner, editors, *Innovative Modellierung und Optimierung von Energiesystemen*, volume 26 of *Umwelt und Ressourcenökonomik*. LIT Verlag, 2009.

[Enq02] Bericht der Enquete-Kommission: Nachhaltige Energieversorgung unter den Bedingungen der Globalisierung und Liberalisierung. Referat Öffentlichkeitsarbeit, 2002.

[Epe07] A. Epe. *Personal communication and documents.* Ruhr-Universität Bochum, 2007.

[ES05] L.F. Escudero and J. Salmeron. On a fix-and-relax framework for a class of project scheduling problems. *Annals of Operations Research*, 140:163–188, 2005.

[Fis81] M.L. Fisher. The Lagrangian relaxation method for solving integer programming problems. *Management Sience*, 27(1):1–18, 1981.

[Geo74] A.M. Geoffrion. Lagrangian relaxation for integer programming. *Mathematical Programming*, 2:82–114, 1974.

[GJ00] E. Gawrilow and M. Joswig. POLYMAKE: A Framework for Analyzing Convex Polytopes. In G. Kalai and G.M. Ziegler, editors, *Polytopes – Combinatorics and Computation*, volume 29 of *DMV Seminar*, pages 43–74. Birkhäuser Verlag, 2000.

[GKKN$^+$02] N. Gröwe-Kuska, K.C. Kiwiel, M.P. Nowak, W. Römisch, and I. Wegner. *Power management in a hydro-thermal system under uncertainty by Lagrangian relaxation*, volume 128 of *Decision Making under Uncertainty: Energy and Power, IMA Volumes in Mathematics and its Applications*, pages 39–70. Springer, 2002.

[GL06] W. Glankwamdee and J. Linderoth. Lookahead branching for mixed integer programming. Technical report, Lehigh University, 2006.

[GMMS09] B. Geißler, A. Martin, A. Morsi, and L. Schewe. Using piecewise linear functions for solving MINLPs. Submitted to IMA Volume on MINLP, 2009.

[GMN$^+$99] R. Gollmer, A. Möller, M.P. Nowak, W. Römisch, and R. Schultz. Primal and dual methods for unit commitment in hydro-thermal power systems. In *Proceedings of the 13th*

Power Systems Computation Conference, volume 2, pages 724–730, 1999.

[GNRS00] R. Gollmer, M.P. Nowak, W. Römisch, and R. Schultz. Unit commitment in power generation - a basic model and some extensions. *Annals of Operations Research*, 96(1-4):167–189, 2000.

[Gri07] V. Grimm. *Einbindung von Speichern für erneuerbare Energien in die Kraftwerkseinsatzplanung - Einfluss auf die Strompreise der Spitzenlast.* PhD thesis, Ruhr-Universität Bochum, 2007.

[HDS07] K. Heuck, K.-D. Dettmann, and D. Schulz. *Elektrische Energieversorgung.* Friedr. Vieweg & SohnVerlag, 7th edition, 2007.

[HH97] S.-J. Huang and C.-L. Huang. Application of genetic-based neural networks to thermal unit commitment. *IEEE Transactions on Power Systems*, 12(2):654–660, 1997.

[HNNS06] E. Handschin, F. Neise, H. Neumann, and R. Schultz. Optimal operation of dispersed generation under uncertainty using mathematical programming. *International Journal of Electrical Power & Energy Systems*, 28(9):618–626, 2006.

[HS08] T. Heinze and R. Schultz. A branch-and-bound method for multistage stochastic integer programs with risk objectives. *Optimization*, 57:277–293, 2008.

[KBP96] S.A. Kazarlis, A.G. Bakirtzis, and V. Petridis. A genetic algorithm solution to the unit commitment problem. *IEEE Transactions on Power Systems*, 11(1):83–92, 1996.

[KdFN04] A.B. Keha, I.R. de Farias, and G.L. Nemhauser. Models for representing piecewise linear cost functions. *Operations Research Letters*, 32:44–48, 2004.

[KM77] S. Kundu and J. Misra. A linear tree partitioning algorithm. *SIAM Journal on Computing*, 6(7), 1977.

[KM05] P. Kall and J. Mayer. *Stochastic Linear Programming.* Springer, 2005.

[Koh08] S. Kohler. Wind, Sonne und Biomasse: Erneuerbare Energien als Teil einer Gesamtstrategie. *η[energie]*, 4:24–26, 2008.

[KV07] C. Küchler and S. Vigerske. Decomposition of multistage stochastic programs with recombining scenario trees. *Stochastic Programming E-Print Series (SPEPS)*, 9, 2007.

[Lee88] F.N. Lee. Short-term thermal unit commitment - a new method. *IEEE Transactions on Power Systems*, 3(2):421–428, 1988.

[LG04] W. Leonhard and M. Grobe. Nachhaltige elektrische Energieversorgung mit Windenergie, Biomasse und Pumpspeicher. *ew*, 103(5):26–31, 2004.

[LL93] G. Laporte and F.V. Louveaux. The integer L-shaped method for stochastic integer programs with complete recourse. *Operations Research Letters*, 13:133–142, 1993.

[LLM04] J. Lee, J. Leung, and F. Margot. Min-up/min-down polytopes. *Discrete Optimization*, 1:77–85, 2004.

[LS99] J.T. Linderoth and M.W.P. Savelsbergh. A computational study of search strategies for mixed integer programming. *INFORMS Journal on Computing*, 11(2):173–187, 1999.

[LS04] G. Lulli and S. Sen. A branch-and-price algorithm for multi-stage stochastic integer programming with application to stochastic batch-sizing problems. *Management Science*, 50:786–796, 2004.

[LW96] A. Løkketangen and D.L. Woodruff. Progressive hedging and tabu search applied to mixed integer (0,1) multi-stage stochastic programming. *Journal of Heuristics*, 2:111–128, 1996.

[Mar05] P. Marcinkowski. Schaltbedingungen bei der Optimierung von Gasnetzen: Polyedrische Untersuchungen und Schnittebenen. Master's thesis, Technische Universität Darmstadt, 2005.

[MK77] J.A. Muckstadt and S.A. König. An application of Lagrangian relaxation to scheduling in power-generation systems. *Operations Research*, 25:387–401, 1977.

[MMM06] A. Martin, M. Möller, and S. Moritz. Mixed integer models for the stationary case of gas network optimization. *Mathematical Programming*, 105:563–582, 2006.

[MMM09] D. Mahlke, A. Martin, and S. Moritz. A mixed integer approach for the time-dependent gas network optimization. *Optimization Methods and Software*, 2009.

[Mor07] S. Moritz. *A Mixed Integer Approach for the Transient Case of Gas Network Optimization.* PhD thesis, Technische Universität Darmstadt, 2007.

[NR00] M.P. Nowak and W. Römisch. Stochastic lagrangian relaxation applied to power scheduling in a hydro-thermal system under uncertainty. *Annals of Operations Research*, 100:251–272, 2000.

[NW88] G.L. Nemhauser and L.A. Wolsey. *Integer and Combinatorial Optimization.* Wiley, 1988.

[Pad00] M. Padberg. Approximating separable nonlinear functions via mixed zero-one programs. *Operations Research Letters*, 27:1–5, 2000.

[PW06] Y. Pochet and L.A. Wolsey. *Production Planning by Mixed Integer Programming.* Springer, 2006.

[Ric08] M. Richter. Relax & Fix Heuristik für ein stochastisches Problem aus der regenerativen Energieversorgung. Master's thesis, Technische Universität Darmstadt, Oktober 2008.

[RS01] W. Römisch and R. Schultz. *Multistage Stochastic Integer Programs: An Introduction*, pages 579–598. Online Optimization of Large Scale Systems. Springer, 2001.

[Sch03] R. Schultz. Stochastic programming with integer variables. *Mathematical Programming*, 97:285–309, 2003.

[SK98] S. Sen and D.P. Kothari. Optimal thermal generating unit commitment: a review. *Electrical Power & Energy Systems*, 20(7):443–451, 1998.

[SLPS90] C. De Simone, M. Lucertini, S. Pallottino, and B. Simeone. Fair dissections of spiders, worms, and caterpillars. *Networks*, 20:323–344, 1990.

[TBL96] S. Takriti, J.R. Birge, and E. Long. A stochastic model for the unit commitment problem. *IEEE Transactions on Power Systems*, 11(3):1497–1508, 1996.

[VAN09] J.P. Vielma, S. Ahmed, and G.L. Nemhauser. Mixed-integer models for nonseparable piecewise linear optimization: Unifying framework and extensions. 2009.

[Web06] C. Weber. Strompreismodellierung - Berücksichtigung empirischer Verteilungen für nicht-speicherbare Güter am Beispiel von Elektrizität. *Essener Unikate*, 29:89–97, 2006.

[Wil98] D.L. Wilson. *Polyhedral Methods for Piecewise-Linear Functions.* PhD thesis, University of Kentucky, 1998.

[Wol98] L.A. Wolsey. *Integer Programming.* Wiley and Sons, 1998.

[Wol08] O. Woll. *Personal communication and documents.* Universität Duisburg-Essen, 2008.

[ZG90] F. Zhuang and F.D. Galiana. Unit commitment by simulated annealing. *IEEE Transactions on Power Systems*, 5(1):311–318, 1990.

MIX
Papier aus verantwortungsvollen Quellen
Paper from responsible sources
FSC® C105338

If you have any concerns about our products,
you can contact us on
ProductSafety@springernature.com

In case Publisher is established outside the EU,
the EU authorized representative is:
Springer Nature Customer Service Center GmbH
Europaplatz 3, 69115 Heidelberg, Germany

Printed by Libri Plureos GmbH
in Hamburg, Germany